Springer Series in Optical Sciences   Volume 19

Edited by David L. MacAdam

Springer Series in Optical Sciences

George A. Agoston

Color Theory
and Its Application
in Art and Design

Second Completely Revised and Updated Edition

With 139 Figures and 23 Color Plates

Springer-Verlag Berlin Heidelberg GmbH

GEORGE A. AGOSTON

4 Rue Rambuteau, F-75003 Paris, France

ISBN 978-3-540-17095-2
DOI 10.1007/978-3-540-34734-7

ISBN 978-3-540-34734-7 (eBook)

Library of Congress Cataloging-in-Publication Data. Agoston, George A., 1920-. Color theory and its application in art and design. (Springer series in optical sciences ; v. 19) Bibliography: p. Includes index. 1. Color. I. Title. II. Series. QC495.A32 1987 535.6 86-29685

© Springer-Verlag Berlin Heidelberg 1979 and 1987
Originally published by Springer-Verlag Berlin Heidelberg New York in 1987

Foreword

This book directly addresses a long-felt, unsatisfied need of modern color science – an appreciative and technically sound presentation of the principles and main offerings of colorimetry to artists and designers, written by one of them.

With his unique blend of training and experience in engineering, with his lifelong interest and, latterly, career in art and art education, Dr. Agoston is unusually well prepared to convey the message of color science to art and design. His book fulfills the hopes I had when I first heard about him and his book.

I foresee important and long-lasting impacts of this book, analogous to those of the epoch-making writings by earlier artist-scientists, such as Leonardo, Chevreul, Munsell, and Pope.

Nearly all persons who have contributed to color science, recently as well as formerly, were attracted to the study of color by color in art. Use of objective or scientific methods did not result from any cold, detached attitude, but from the inherent difficulties of the problems concerning color and its use, by which they were intrigued. Modern education and experience has taught many people how to tackle difficult problems by use of scientific methods. Therefore – color science.

Few artists or others who deal with color will deny that color poses difficult problems. Capable people, all well-disposed to art and the aesthetic approach, have recently added significantly to the knowledge of color and to ways of working with it. They always intended that their findings would be useful to artists and designers. Unfortunately, they have not succeeded in conveying that message, or their contributions, to those intended beneficiaries. This book by Dr. Agoston will, I think, be the bridge of color between the cultures of science and art, of which modern color scientists have dreamed but never succeeded in building.

The book is understandable by all persons, no matter what their education or experience. Everyone is interested in color. This book has a lot for everyone, no matter how little they have to do with color, nor how little their acquaintance with or interest in mathematics or physics. No equations are used. There are many graphs. They can be understood by anyone who reads newspapers or news magazines. Each graph, and its meaning for color, is explained in simple words. Yet the book is not condescending or trivial.

Knowledgeable scientists will find facts and perspectives that are not found elsewhere, some of which will be new and stimulating, even to color scientists.

Rochester, September 1979   *David L. MacAdam*

Preface to the Second Edition

This new edition includes some updating, the expansion of the discussions of some topics, and the addition of new topics and illustrations. My aim is unchanged. I wish to make selected pertinent knowledge in color science accessible to artists and designers. I continue to adhere to the policy of avoiding the inclusion of mathematical equations and texts for which a technical background of the reader must be assumed. But, for the benefit of those who do possess a technical background, I have introduced a Notes Section (Appendix) in which some equations and supporting technical information are presented. The material in the Notes Section is supplementary; it is not required reading for a full comprehension of the main text.

Chapter 11, entitled "Conditions of Viewing and the Colors Seen", is the major addition to the book. Here several selected topics in psychology and physiology are discussed that extend the scope of the preceding chapters. The topics are clearly within the domain of interest in art and design. They include: color response in vision, adaptation, afterimages, simultaneous contrast, colored shadows, edge contrast, and assimilation. The discussion of the physiology of the eye in Chap. 2 has been expanded somewhat in preparation for the material in Chap. 11.

Much of the discussion of afterimages in Chap. 11 deals with afterimage complementary pairs. This is the type of complementary colors considered by Goethe in his color studies and frequently employed by artists in the past. There are many discussions of afterimages in the old scientific color literature. The more recent study by M.H. Wilson and R.W. Brocklebank (1955) stands out as one that offers potentially useful information on complementary pairs. The discussion of the Goethe color circle in the first edition has been modified here to take afterimage complementary pairs into account.

The discovery of the attribute brilliance by the color scientist R.M. Evans, referred to in Chap. 3 of the first edition, does not seem to have stirred up much interest among those currently engaged in color vision research. Yet, in my view, it is a potentially useful concept for artists and designers. More recently, R.W.G. Hunt has considered in detail the attributes of perceived color and, in particular, their definitions. Both what he calls colorfulness and the brilliance identified by Evans are treated in this edition.

Artists and designers are becoming increasingly knowledgeable about the utility of the CIE chromaticity diagram. The CIE 1931 chromaticity diagram is the dominant tool employed in color specification today. However, increasing use is being made of the CIE 1976 chromaticity diagram, which is an approximately uniform version. A discussion has been introduced in which the latter is presented.

The text concerning the especially important OSA Uniform Color Scales has been expanded and now occupies a chapter by itself (Chap. 9). Diagrams are provided to identify both uniform color scales and uniform two-dimensional color arrays in the OSA scheme. The CIELUV and CIELAB color spaces possess similar potential value in art and design. These are both discussed in greater detail (Chap. 8).

A number of sections in Chaps. 7 and 8 have been expanded, and the following new topics have been introduced: iridescent colors (liquid crystals), metameric illumination, color rendering, the German Standard Color Chart, and two Japanese color sample sets (*Chroma Cosmos 5000, Chromaton 707*). The change in the Swedish NCS scheme of notation was not included in the earlier edition of this book; the revised notation is presented here.

In the Introduction of the earlier edition, I cited names of contemporary color scientists who have made technical contributions in areas that are particularly pertinent in art and design. Not surprisingly, I have since learned of other major contributors, and, if I attempted a new list, I fear that it, too, would be inadequate. Now I appreciate better the rather widespread concern in color science for the needs of artists and designers.

In the preparations for this new edition, I have had the help and support of a number of people. Once more I am particularly indebted to Dr. David L. MacAdam for his critical comments. I am thankful to him for providing the color photograph for Plate XII and the color samples used in preparing Plate XIII. I am grateful to Prof. Leo M. Hurvich for his attention to my numerous inquiries concerning his research, and to Miss Dorothy Nickerson and Dr. Fred W. Billmeyer, Jr. for their assistance on various occasions. In this edition, I have made use of information that had been generously offered six years ago by Mr. Kenneth L. Kelly, of the National Bureau of Standards, Washington, D.C., some of which I was unable to include in the first edition.

I wish to acknowledge with thanks the help of French artist Yves Charnay who supplied the two photographs of his painting in liquid crystals for use in Plate III; of Dr. Takashi Hosono, President of the Japan Color Institute, Tokyo, who provided information about Chroma Cosmos 5000 and Chromaton 707, and supplied the photograph for Plate XI and the diagram for Fig. 8.26; of Mr. Rolf G. Kuehni, Mobay Chemical Corp., Rock Hill, South Carolina, who provided the spectral reflectance curve for Fig. 7.19; and of Munsell Color, Baltimore, who provided photographs for Fig. 8.11 and Plate VIII and a set of CIE/Munsell conversion charts, which were used for the preparation of Figs. 12.1–9. I wish to thank, too, Dr. Nahum Joël for continued helpful discussions in the domain of physics. Again, the

Documentation Services of the Eastman Kodak Co. at Vincennes, France, have aided by giving me access to their reference materials, and I am grateful for this help.

These and others have contributed in various ways to my book. However, the decisions that I have had to make in writing the book are my own, and I accept the responsibility for what I have written.

I appreciate, in particular, Marjorie's tolerance. When she married me, I was fully occupied as a painter. She certainly never dreamed that one day I would take on the demanding task of writing a book on color.

Paris, March 1987 *George A. Agoston*

Preface to the First Edition

My aim in this introductory text is to present a comprehensible discussion of certain technical topics and recent developments in color science that I believe are of real interest to artists and designers. I treat a number of applications of this knowledge, for example in the selection and use of colorants (pigments and dyes) and light.

Early in the book I discuss what color is and what its characteristics are. This is followed by a chapter on pertinent aspects of light, light as the stimulus that causes the perception of color. Then the subject of the colors of opaque and transparent, nonfluorescent and fluorescent materials is taken up. There are sections on color matching, color mixture, and color primaries. Chapter 6 introduces the basic ideas that underlie the universal method (CIE) of color specification. Later chapters show how these ideas have been extended to serve other purposes such as systematic color naming, determining complementary colors, mixing colored lights, and demonstrating the limitations of color gamuts of colorants. The Munsell and the Ostwald color systems and the Natural Colour System (Sweden) are explained, and the new Uniform Color Scales (Optical Society of America) are described.

Color specification itself is a broad topic. The information presented here is relevant in art and design, for those who work with pigments and dyes or with products that contain them, such as paints, printing inks, plastics, glasses, mosaic tesserae, etc., and for those who use colored lights, lasers, and phosphors. I believe that this book can be of use as an introductory text to others in art conservation and in industries and commerce concerned with printing, dyeing, plastics manufacture, etc., but I have not treated their particular technical problems and have not introduced their specialized terminologies.

I have taken great care to present technical information in a simple yet undistorted manner. No background in science or mathematics is necessary to follow the text. Algebra is not employed, but graphs, which are indispensable in discussing the subjects, are used in an elementary way. I believe that readers who are familiar with graphical presentations of the sort found in daily newspapers and news magazines will have no difficulty in understanding the graphs in the book. (...)

The text is based on information drawn principally from the current technical literature in color science, a domain that is found by many to be forbidding, especially because it extends into rather different scientific disciplines, principally psychology, physiology, and physics. Numbers within brackets, such as [2.4] and [8.26], indicate citations of books and articles listed in the References Section at the end of the book. The notation [5.6, 7] signifies Refs. 5.6 and 5.7. In some cases a further distinction is made by giving the page number of a book, as [Ref. 6.2, p. 171].

My interest in color is that of an artist. It had its start in my early teens when I began making oil paintings. Then there was a gap in my art career when I studied and worked twenty years as a chemical engineer. Later, when I returned actively to painting, I came under the influence of artist and teacher Richard Bowman, and my use of color in painting changed radically from realistic to fauve. The development of my interest in the technical aspects of artists' materials and in the subject of color relates somehow to my training and experience in engineering. However, I can point with certainty to my physicist friend Dr. Arthur Karp as the one who kindled my interest in the basic topic of color perception.

I am indebted to Dr. David L. MacAdam for his critical reading of the manuscript, to Dr. Nahum Joël for his helpful comments on the first half of the text, and to Mr. Kenneth L. Kelly for his suggestions concerning sections of the text dealing with certain work done at the National Bureau of Standards, Washington, D.C. I am thankful to personnel of the Documentation Services of the Eastman Kodak Company at Vincennes, France, for making reference materials available to me.

Paris, September 1979   *George A. Agoston*

Contents

1. Introduction

1.1 Color Science and Art Before 1920

Scientific aspects of the phenomenon of color perception have captured the interest of artists, musicians, and writers during the past two centuries. The German poet Goethe made many detailed observations about color perception and presented his ideas in a book entitled *Farbenlehre* (Theory of Colors) (1810) [1.1,2], which in the opinion of a prominent color authority Deane B. Judd (1900–1972) "may come to be recognized as foreshadowing, however dimly, the next important advance in the theory of color" [Ref. 1.1, p. xvi].

J.M.W. Turner studied Goethe's book on color and produced some compositions based on it [1.3]. His lecture notes at the Royal Academy reveal his interest as well in the work of the scientist-mathematician Isaac Newton on light and color [1.4]. In France, Eugène Delacroix applied principles that he had learned from *De la loi du contraste simultané des couleurs* (The Principles of Harmony and Contrast of Colors) (1839) by Michel-Eugène Chevreul, chemist and director of the dye houses of the Gobelin Tapestry Works outside (now inside) Paris [1.5,6]. Neo-impressionists Georges Seurat and Paul Signac were profoundly influenced by the book *Modern Chromatics* (1879) by the American artist-physicist Ogden Nicholas Rood and applied their knowledge in their divisionist paintings [1.5,7]. In recent years, new interest in Chevreul's book has been stimulated by the artist Josef Albers (1888–1976) at Yale University [1.8–10] and by the work of Op artists who have sought ways to heighten color brilliance.

A.H. Munsell (1858–1918), artist and teacher at the Massachusetts Normal Art School (now the Massachusetts College of Art) (Boston), was particularly interested in finding an appropriate method for teaching color to children [1.11]. He devised a practical color-notation system that had a scientific basis to serve as a teaching aid. Within several decades his system assumed great importance in color science and in color technology. In 1905 Munsell complained of "the incongruence and bizarre nature of our present color names" [1.12]. Pointing out that "music is equipped with a system by which it defines each sound in terms of its pitch, intensity, and duration," he reasoned that color should "be supplied with an appropriate system based on the hue, value, and chroma of our sensations ... ." The Munsell color system now serves as one important means for color specification. Other roles have been found for it. Munsell himself proposed how it could be used in choosing harmonious colors [Refs. 1.13, p. 129; 1.14]. Color harmony has been discussed by Goethe [1.1], Chevreul [1.6], Rood [1.7], and the chemist Wilhelm Ostwald (1853–1932) [1.15–17]. The subject of color harmony has been treated in a book on colorimetry by Judd and Wyszecki [Ref. 1.18, p. 390], more generally by Judd in a well-illustrated booklet [1.19], and analytically by Burnham, Hanes and Bartleson [Ref. 1.20, p. 214].

1.2 Some Developments in Color Science Pertinent to Art and Design Since 1920

Denman Ross (1853–1935) and Arthur Pope (1880–1977) introduced color theory to their art and design students at Harvard University (Cambridge, Massachusetts) more than fifty years ago [1.21,22]. In that early period, Byron Culver (1894–1971) also presented the same subject at the Department of Applied Art of the Rochester Atheneum and Mechanics Institute (now the Rochester Institute of Technology) (Rochester, New York) [1.11]. Similar courses have been offered at many other art schools and departments of art and design of universities. But the examples set by Ross, Pope, Culver, and undoubtedly others were evidently exceptions. In 1942, R.B. Farnum of the Rhode Island School of Design reported, following a survey, that sometimes such subjects were treated only incidentally and that too little time was allotted to them. Some entrusted to teach color theory were incompetent or insufficiently interested [1.23]. Today, undoubtedly because of the impact of new developments in science and technology, more art schools and departments of art and design are giving fuller attention to the teaching of pertinent topics in the area of color science. Art teachers and artists are writing articles about their applications of color theory and their color research [1.24–31]. The international art journal *Leonardo*, which treats contemporary visual art, with full recognition given to pertinent aspects of science and technology, has presented a number of diverse articles on the subject of color.

In commerce and industry, much attention has been given to color specification. For this purpose, Munsell implemented his color system with a large set of very carefully prepared color samples. The samples were related to one another through progressive changes of approximately equal steps of Hue, Value, and Chroma. For diverse applications, a number of other sample systems have been devised that are charcterized by other features. Common to most such standardized systems, each of which has hundreds of samples, is the practice of assigning numbers or codes to the colors.

Thus, colors matched to a standard sample are identified precisely by the corresponding number or code. This procedure is useful for communication, in commerce for example, where the use of color samples themselves would be inconvenient.

An internationally accepted method developed by the Commission Internationale d'Eclairage (CIE) is widely employed for specifying color. It is based on the fact that the relative amounts of three standard primary colors required in a mixture to match a color can be used to identify and specify the color.

The CIE method has been applied in subsidiary ways as well, some of which are of particular interest to artists and designers. These applications refer to the simple graphical presentation that serves the CIE method. The graphical presentation provides a basis for selecting, for example, the color names for lights. It enables the prediction of the colors obtainable when two or more lights of known color are mixed. In another application, the change of color quality (hue and purity) is traced when paints are mixed and when the color of a paint film fades with time. The graphical presentation also provides a basis for selecting additive complementary colors. Also, the upper purity limits for colors of nonfluorescent pigments and dyes can be shown on the graph, for comparison with the purities obtained with presently available paints and inks. Furthermore, the CIE scheme is a stepping stone to other schemes that provide for precise determination of color differences. This is of particular interest to those concerned with close control of color differences in their work and to those wishing to know specifically about the precise degree of color change (as in a fading or deterioration of pigments).

Both the Munsell and the Ostwald color systems have been known to artists and designers for a long time. An adaptation of the latter system is represented by a collection of samples in the *Color Harmony Manual* [1.32], a collection provided primarily for use in design. A more recently introduced sample collection, the Swedish *SIS Colour Atlas NCS* (*Natural Colour System*) (1979) [1.33], will probably be of great importance to designers, artists, and architects. The NCS, like the Munsell color system [1.34], provides color samples selected by visual means. Of great significance is the fact that anyone with normal vision can apply the NCS method of color judgment without the use of samples and color-measuring instruments. In 1977 a collection of samples was made available by the Optical Society of America, which provides many series of colors of equal color difference [1.35]. The collection has been produced for applications in art and design as well as for study in color science. In Japan a number of collections of color samples have been produced. The most outstanding ones for use in design are *Chroma Cosmos 5000* (1978) and *Chromaton 707* (1982) [1.36,37].

In the English language, there is a profusion of names for colors in art, science, and commerce. Many names apply to more than one color, and many colors are labelled by more than one name. In an attempt to establish some order, one major effort has been made by the U.S. National Bureau of Standards (NBS) and the Inter-Society Color Council (ISCC) to produce and identify a set of about 300 easily recognized and consistent color names and to provide a dictionary that relates over 7000 currently employed color terms to the set. Thus, for example, the term "Hooker's green", familiar to many artists, but not all, can with the aid of the dictionary be replaced by the more universally recognizable terms "strong yellowish green" or "dark yellowish green", depending on Munsell Value and Chroma. The ISCC-NBS color names have been adopted by *Webster's Third New International Dictionary* [1.38] and are in wide use in commerce. However, artists and designers, who were also expected by the originators of the color-name system to derive direct benefit from it, often seem to be unaware of it.

The advances in color science mentioned in the preceding paragraphs concern the rather specialized domain of colorimetry. By comparison, the subject of color vision is vast. It embraces several conceptual levels [Ref. 1.39, p. 3]: anatomical (the path from eye to brain), physiological (chemical and electrical mechanisms), physical (the passage of light from source to retina), psychological (the gross response of the visual system), and esthetic (pleasure, well-being). In spite of the enormous amount of research that has been reported, many basic underlying facts have yet to be revealed before a complete description or overall theory of color vision can be devised.

Some scientific topics in the psychological domain that are of real interest in art and design are chromatic adaptation, color constancy, afterimages, simultaneous contrast, and assimilation. Many artists are familiar with these either from their formal studies or from practice. These topics are of fundamental importance in color science and are being actively studied. They are discussed in this book along with pertinent topics in colorimetry, physics, and anatomy. The important domain denoted by the word "esthetic" involves both psychology and philosophy; it is not treated in this book. Nevertheless, scientists are interested in it, as is suggested by their writings on color harmony (above).

The fact that there are aspects of color science that are of practical interest to artists and designers has been recognized by color experts for a long time. In recent decades, many have contributed to color science in areas that are particularly pertinent to art and design. Their work has already made an impact on art conservation as practiced in museum laboratories.

It seems ironical that, although students and professional artists are rather well acquainted with earlier developments, such as the Munsell and the Ostwald color systems, many are unfamiliar with the comparatively recent strides in color science that are not only available to them but are also intended, in part, for their use. I hope that this book will help to arouse their interest in this new knowledge.

4

2. The Concept of Color

2.1 What Is Color? One Answer

The color authority R.M. Evans (1905–1974) pointed out that the word *color* "as it is used in ordinary speech ... has many different meanings" [Ref. 2.1, p. 173]. Even in the scientific domains of chemistry, physics, and psychology it has different specialized meanings.

Let us consider first an everyday usage of the word *color*, a usage that implies the concept that color is a *property of materials*. Thus a ripe tomato has the property of being red; snow, of being white; Mary's scarf, of being blue; etc. We accept that the color of a material or object is its color perceived in daylight.

We also speak of the color of light and commonly consider color to be a *property of light*. Thus when we look at a red traffic light, we suppose that light of a red color is radiating to our eyes. The red disk produced when such a beam of light strikes a white wall seems to confirm that the light has a red color.

The concepts of color as a property of materials and of light serve numerous practical needs in daily life, the most important of which are those of survival [Ref. 1.39, p. 7]. The concepts serve well in a host of ways in commerce, art, science, and technology. For these reasons, it comes as a surprise to many people that these concepts are *incorrect*. Perceived color is incorrectly given as a property of materials and as a property of light. Newton, discussing the subject of light in his book *Opticks* (1704), stated correctly, "Indeed, rays, properly expressed, are not coloured" [Ref. 1.1, p. vii].

Another example of an incorrect concept that serves us well in daily life is the idea that the sun rises and sets every day. Objective observations reveal, however, that the sun is not revolving about the earth, producing a sunrise and a sunset every 24 hours; instead, the spinning of the earth on its axis (one revolution every 24 hours) cuts off our view of the sun at the moment we call sunset and permits our view of it again at the moment we call sunrise.

The concepts of color as a property of materials and of light and the concept that the sun rises and sets every day will continue to serve us in many important ways. We should, however, be prepared for the frequent instances when the concepts of color as a property do not apply. Here is one example: The colors of two different fabrics match in one illumination but not in another. If color is a property, why does the color change and why does the match not remain? (We shall see later in this book that *metamerism* is involved.) Here is another example: Using one green paint, an artist makes a circular opaque disk (about 3 cm in diameter) in the center of each of two sheets of paper (20 × 20 cm); one sheet is colored dull red and the other neutral gray, but both the red and gray have the same Value (Sect. 8.4). The disks will not be perceived to have the same green hue. Why does the color of the disk change when the color of its background is changed? (The visual phenomenon occurring here is called *simultaneous contrast*.)

2.2 The Visual System: A Brief Sketch

Colors may be perceived with the eyes closed, in dreams [Ref. 2.2, p. 176], and also when physical pressure is applied to the eyeballs or when certain drugs have been taken [Refs. 1.39, p. 6; 2.3, p. 14]. (Obviously, these are situations where color is not a property of materials or a property of light.) With the eyes open, under the usual conditions of seeing colors, light enters each eye and becomes absorbed in its interior; a succession of events follows that leads to the production of a signal or sensation in the brain. The sensation makes us aware of a color. After stating that light "rays ... are not coloured," Newton wrote, "There is nothing else in them but a certain power or disposition which so conditions them that they produce in us the sensation of this or that colour" [Ref. 1.1, p. vii]. But the awareness of color does not tell us what color is. Before we continue with the question, What is color? (Sect. 2.3), however, it will be helpful to consider briefly certain information concerning the visual system.

When we look at a *luminous object*, such as a piece of red-hot iron, a glowing incandescent-lamp filament, or the sun, our eyes react to light radiating directly from the object and we see it. On the other hand, a *non-luminous object*, such as a tree or a table, must be illuminated to be seen – that is, if light (in sufficient amount and of adequate quality) falls upon the object, some of the light will be diffusely reflected (*scattered*) and our eyes will react to the reflected light [Ref. 2.1, p. 54]. Out of doors, nonluminous objects may receive light in beams radiating directly from the sun, but in general they receive sunlight indirectly, scattered by clouds[1], clear sky [1] and

---

[1] Light from clouds is sunlight scattered by water droplets. Dust particles and molecules (of oxygen and nitrogen, principally) in the atmosphere also scatter sunlight. The scattering by gas molecules is wavelength dependent, producing the blue color of the sky [2.4].

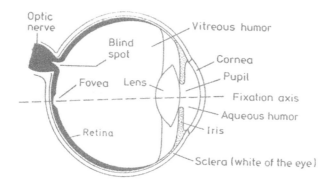

Fig. 2.1. Schematic diagram of the horizontal cross section of the right eye (top view)

surrounding nonluminous objects (foliage, walls, etc.). Indoors, during the daytime, scattered light and sometimes sunbeams enter through windows. At night, nonluminous objects often receive directly radiated light and diffused light from lamps and, importantly for surfaces in shadow, scattered light from surrounding nonluminous objects (furniture, walls, etc.).

Light reaching an eye enters it through a clear liquid (*aqueous humor*) and passes through the *pupil* and the *crystalline lens*. It then traverses a bulk of jellylike material (*vitreous humor*) and falls upon and is absorbed by the *retina*, a thin covering extending over the rear inner surface of the eye (Fig. 2.1) [Refs. 1.18, p. 6; 1.20, p. 44; 2.2, p. 7; 2.5, p. 16].

The retina consists of 10 layers identified as nerve cells, nerve connections, and membranes [Refs. 2.2, Fig. 2.3; 2.3, p. 114; 2.6, Fig. 3]. The light must pass through at least five layers before it arrives at the light-sensitive photopigment-containing *receptor cells*, where light is absorbed. The absorption of light by a cell results in a reversible chemical change in the pigment. The chemical change initiates a transformation involving an electrical change that, in turn, somehow influences certain cells closer to the surface of the retina in which spike electrical discharges are produced. This excitation is transmitted through optic nerve fibers from each eye to the opposite half of the brain, to a part called the lateral geniculate nucleus (LGN), and from there on to visual centers at the rear of the brain.

In the brain, sensations are produced that reveal aspects of appearance of the objects in view. Several such aspects are size, position, color, glossiness, texture, opacity, and transparency. The fact that there are some interconnections between juxtaposed nerve cells, which permit light that falls on one part of the retina to affect what is seen in another part, is cited as an explanation of visual phenomena such as simultaneous color contrast [2.6].

There are two types of light-sensitive receptor cells in the retina, known as *rods* and *cones* (because of their shapes) [Ref. 2.3, p. 116]. The rods, which respond only to light and dark, are characterized by high sensitivity; they are capable of responding to light of very low intensity. The rods enable us to see in dimly lit rooms or in moonlight. At such low levels of illumination we are unable to distinguish hues (reds, yellows, greens, etc.) and we cannot discern detail as well as we can in daylight [Ref. 2.5, p. 17]. At very low light intensities the cones, which are responsible for color vision, are considered not sensitive enough to respond. Our vision at very low light intensities, essentially without color response, is called *scoptic vision* [Note 2.1].

When the level of illumination is sufficiently high – that is, within the range of illuminance produced by street lighting and bright sunlight [Ref. 2.7, Fig. 2] – the cone system responds. It is presumed that the rod system is then more or less saturated and hence incapable of producing significant variations in response [Ref. 2.8, p. 569]. We then perceive colors and experience *photopic vision*. There are three classes of cones. In recent versions of the opponent-color theory of color vision, the three classes of cones are red-, green-, and blue-producing. It is assumed that, by appropriate interactions involving two or three classes, all the achromatic and chromatic colors, over the full range of brightness level (in photopic vision), can be produced (grays, red yellows, yellow reds, red blues, blue reds, green yellows, yellow greens, green blues, blue greens).

When an eye is focused on an object, its image falls on (or across) a small depression in the retinal surface called the *fovea* (Fig. 2.1). This region differs from the surrounding retinal area in that the layers through which light must travel before reaching the receptor cells are thinner. The receptor cells in the central region of the fovea (subtending an angle of vision of 1°) are all cones (about 25 000 of them!). The cones are more densely packed and longer than cones elsewhere. In the surrounding area of the fovea (subtending an angle of 2° or more), some rods are present, but they are relatively few at angles less than about 10°. Hence the fovea's center is the region of highest visual acuity in photopic vision [Ref. 2.2, p. 9]. The fovea determines the fixation axis when we focus our eyes on a detail (Fig. 2.1).

There is a relatively small range of illuminance at which both rods and cones respond significantly and contribute to what is seen. Then "twilight vision" (*mesopic vision*) is experienced [Ref. 2.5, p. 18].

2.3 What Is Color? Some Other Answers

Now let us return to the question, What is color?. The psychologist L.M. Hurvich poses this question in his book [Ref. 2.3, p. 13]. He asks whether an object has color because of its physical-chemical makeup or whether illumination constitutes the color of the object. Continuing, he asks whether color is a photochemical event in the retina, a neural brain-excitation process, or a psychical event. His answer is, "Color is all these things ...," but he adds that, before exploring these topics, "the main point to be made is that our perception of color ordinarily derives from an interaction between physical light rays and the visual system of the living organism. Both are involved in seeing objects and perceiving color."

Kuehni devotes the first chapter of his book to the question, What is color?; in it he discusses physical, physiological, psychological, and psychophysical aspects. In answer to the question, he proposes a psychologically oriented definition: "Color ... is an experience, poetically speaking a flower of our brain activity" [Ref. 1.39, p. 7]. He deplores the official proposal of the following "circular" definition of color for scientific use. But I quote it here because it does convey a related psychological meaning: "Perceived color is the attribute of visual perception that can be described by color names: White, Gray, Black, Yellow, Orange, Brown, Red, Green, Blue, Purple, and so on or by combinations of such names."

In his book on color, Evans introduced the question in a way similar to that of Kuehni. In much the same way as Hurvich, he wrote, "Any attempt to arrive at a definition of the word involves one at once in all the complexities of vision" [Ref. 2.1, p. 1].

In discussions of color perception, the term *color stimulus* (or simply *stimulus*) is generally used to refer to the light that arrives at the retina. Perception of a color by the brain is designated by the term *color response* (or simply *response*). The words "stimulus" and "response" are used later in this book where reference is made to color perception.

2.4 What Is Color? A Practical Answer in Technology

Color scientists and technologists interested in quantitative means for specifying and measuring color have avoided the "complexities of vision" by defining color to be a characteristic of light, the *stimulus*. Indeed, color is a topic in the domain of psychophysics where a quantitative scheme has been devised – namely, the CIE-system, discussed in Chap. 6.

To obtain an idea of the rationale of the psychophysical approach, let us consider the following hypothesis: Light enters the eye and is absorbed by the retina. A series of events is caused to occur that lead to the production of a signal or sensation in the brain. The sensation makes us aware of a characteristic of the light. Color is this characteristic. (Note that color is not a sensation.) Alternatively and equivalently, color is the characteristic of materials that results in their changing the characteristic of the illuminating light. Thus, the red color of a traffic light is a characteristic of the light; the green color of a leaf is a characteristic of the leaf that produces a change in the characteristic of, say, the daylight in which it is found. (These same sensations, which are normally caused by light, can also lead to illusions of color when the eyes are closed, as in dreams or when pressure is applied to the eyeballs.)

According to this hypothesis, the characteristic (color) of the light is its spectral power distribution (wavelength composition); the characteristic (color) of an opaque material is its spectral reflectance distribution, considered along with the characteristic (color) of the light illuminating the material (Sects. 4.5,6; 5.2). Thus, color, characterized in this way, is given by numerical data or curves representing the data.

In science and technology, the color of light and of materials is commonly characterized by such curves. In addition, by means of a CIE method, such data can be combined with quantitative information that specifies the sensitivity of a typical human eye, to calculate a color designation (Chap. 6). The CIE numerical color designations are particularly useful because they can be identified with colors seen under standardized viewing conditions.

The concepts of color as a characteristic of light and of materials bear a strong resemblance to the concepts of color as a property of materials and of light. These concepts refer to a stimulus. The difference is that, in the former concepts, color is a set of data and, in the latter, it is a subjective judgment of an aspect of appearance.

3. Perceived Colors

3.1 Isolated Colors

When we focus our eyes on a uniformly colored area of a painting, the color that we perceive is often influenced by the colors of surrounding areas. In the preceding chapter, it was mentioned that this psychological phenomenon is called simultaneous contrast. Artists and designers deal with it in striving for specific color effects.

If, on the other hand, we wish to discuss or specify the precise color of a paint sample, for the sake of simplicity the sample should be considered in isolation, without the influence of colors of the surroundings, or in a standardized situation such as with a white or neutral gray background.

When we see a red railway signal glowing from a distance at night in the absence of other lights, we are experiencing an *isolated* or *unrelated color*. The light received solely from one such source is called an *isolated stimulus*. Often the situation of an isolated stimulus is closely approached when the surroundings are not black if the intensity of the light (stimulus) greatly exceeds that of all of the surroundings.

Usually it is not difficult to devise a way to receive light in isolation from a luminous object. But how can the light that is scattered from an object such as a piece of paper or a sample of paint be viewed in isolation? One way is to illuminate the object in an otherwise darkened room. Another way is to view the surface through an aperture or round hole in a black shield (*reduction screen*) while focusing on the perimeter of the hole. (The larger hole in the reduction screen inserted inside the back cover of this book may be used for this purpose.) The view of the uniformly colored surface some distance behind the screen should fill the hole. Because the black shield does not reflect much light, practically all the light received by the eye arrives through the hole from the surface of the viewed object. Also, because the hole's perimeter, not the object's surface, is in focus, the viewer gets the impression of a diffuse filmlike zone. (When the smaller hole is used, it should be held close to the eye.) Such color perceptions are not located in depth [3.1]; they are often called *film colors* or *aperture colors*.

Aspects of the appearance of objects such as glossiness, transparency, and surface texture are eliminated when surfaces are viewed in this way. Ordinarily, these characteristics interfere with the assessment and comparison of surface colors [3.2]. For example, consider the task of selecting a silky fabric to match the color of a woolen fabric.

Color that is perceived to belong to an object (self-luminous, like a lamp filament, or non-self-luminous like a dab of paint or a wine bottle) is called *object color*. The color of a non-self-luminous opaque object is often more specifically referred to as *surface color*. A film color is a *nonobject color* [3.1].

3.2 Hue

Perceived colors have been found by Evans to have as many as five different attributes [Ref. 2.5, p. 94]. More recently, R.W.G. Hunt [3.3] and K. Richter [3.4] have called attention to the existence of several more. In the simplest case, that of isolated colors or film colors, there are just three attributes: *hue, saturation,* and *brightness* [Ref. 2.5, p. 136]. Let us ask first: What is hue?.

When we look at a red light, we perceive a red *hue*. It is difficult to explain just what the perception of a red hue is, just as it is difficult to explain the perception of bitterness or the aural perception of shrillness. It is sufficient for our purposes to say that when we utter or write the word "red", or the words "blue" or "purple", we are conveying to others the idea of a particular hue. It has been estimated that a normal eye can distinguish about 200 hues [Refs. 2.1, p. 118; 3.5].

Perceived colors that possess a hue are called *chromatic colors*; those that do not are called *achromatic colors*. We perceive an achromatic (hueless) color when we look at a glowing daylight fluorescent lamp, for example. We also perceive achromatic colors when we view white, neutral gray, or black surfaces illuminated by such a lamp or by daylight.

It has been found that among all the hues there are only four that are not perceived as mixtures. These are called the *unitary*, or *unique, hues* [Ref. 2.5, p. 66]: *unitary red, unitary yellow, unitary green,* and *unitary blue.* All other hues are considered to be *binary hues* because they are seen as mixtures of the following pairs: unitary green and unitary yellow (yellowish greens and greenish yellows); unitary yellow and unitary red (reddish yellows and oranges); unitary red and unitary blue (magentas, purples, and violets); unitary blue and unitary green (greenish blues and bluish greens):

It was mentioned in Sect. 2.2 that three classes of cones have been identified in the retina. This is consistent with a hypothesis that states that three opponent pairs of *psychological primaries* (white and black, red and green, and yellow and blue) are involved in color vision [Refs. 2.3, p. 17; 2.5, p. 107]. The four unitary hues are basic to this hypothesis. In Sect. 8.7 the use of the four unitary hues and white and black in a practical method for judging colors is described, and in Sects. 11.2,3 their role in hue response and their application in a psychological color specification system are discussed.

3.3 Saturation and Colorfulness

Perceived chromatic colors can generally be considered to possess a hue component and an achromatic component. *Saturation* is an attribute of perceived color according to which we judge the *relative* amount of the hue component in the color [Note 3.1]. Let us consider two isolated beams of light, one red and the other pink. The beams are such that each evokes the same perception of hue and of brightness. The pink has the lower saturation because the relative amount (concentration) of the red component is less. L.M. Hurvich and D. Jameson have measured saturation coefficients quantitatively in their studies of color perception (Sect. 11.3) [Refs. 2.3, p. 79; 3.6,7]. The saturation coefficient is calculated by dividing the amount of the hue component by the sum of the amounts of the hue and achromatic components. It is the fraction or percentage of hue in a color and is considered to be a quantitative measure of perceived saturation.

There is no unanimity among color scientists on the precise definition of saturation [Refs. 1.39, p. 39; 2.5, pp. 119, 184; 3.3]. The above definition is ample for our purposes and is used in this book.

Hunt has coined the term *colorfulness* for the attribute of perceived color according to which we judge the *absolute* amount of hue component in the color, irrespective of the amount of achromatic component present [3.3] [Note 3.2]. (Another term, *chromaticness,* is also used to designate the same attribute [Ref. 1.39, pp. 39, 48]). Thus, pink in the above example would be said to have less red hue than the color of the other beam. In other words, colorfulness applies to the absolute chromatic response experienced.

Here is an example that demonstrates the difference between saturation and colorfulness. A piece of red paper illuminated by the beams from two identical spotlights in a darkened room has less colorfulness when one of the spotlights is turned off; less red hue is perceived. The saturation, however, is unchanged − that is, the concentration of red (the amount of red with respect to the total amount of red and achromatic content) does not change. Similarly, the color of a red dress viewed out of doors exhibits diminished colorfulness (but unchanged saturation) when viewed indoors under reduced illumination of the same quality. As Hunt has pointed out, saturation is the more important attribute of color for the recognition of objects. Colorfulness, on the other hand, depends on the illumination, a fact of real interest in art and design.

3.4 Brightness and Lightness

Evans summarized in a book (1974) his thoughts and experimental evidence concerning the attributes of perceived color [2.5]. His discovery of the attribute brilliance may well be a major contribution to the science of color perception, but to accommodate brilliance among the other known attributes requires an expert reexamination of the respective roles played by each, particularly by saturation and by lightness.

Brightness is an attribute of the illumination in which a nonisolated object is viewed [Ref. 2.5, pp. 96, 123]. Brightness commonly increases when the intensity of illumination increases. More precisely stated, brightness is the "perception of the general luminance level" [Ref. 2.5, p. 93]. (The term "luminance" is considered in Sects. 6.2,3.) Brightness can refer to the perceived color of an object only when the object is isolated and light comes to the eye from the object and from nowhere else. For example, it is permissible to talk about the brightness of the color of light from a lamp or from a piece of paper illuminated by a spotlight observed in an otherwise darkened room. The visual experience of brightness is commonly described at the limits of its range as "dim" and "dazzling".

Perceived lightness is an attribute of nonisolated colors (related colors). A related color would be perceived, for example, while viewing a green vase against a brown panel. Lightness is produced by the presence of a second stimulus or of the surroundings [Ref. 2.5, pp. 136, 137]. It commonly implies comparison, such as "lighter than" or "darker than" something else; it implies a perception of the luminance of light from one area relative to that from another or from the surroundings [Ref. 2.5, p. 93]. We perceive lightness when we sense that more light is coming to our eyes from a piece of paper than from the brown table on which it lies.

Evans has objected to the general practice of linking lightness to brightness in considerations of the colors of nonluminous objects. Thus, a definition beginning as follows would be considered misleading: "The term 'lightness' is used in place of 'brightness' to refer to surfaces ... ." By such a definition, the perceptions of grayness and of darkness are incorrectly linked to brightness [Ref. 2.5, p. 93]. Evans' experimental work showed that brightness and lightness are separate variables, which had also been noted by others [3.3, 3.8]. Furthermore, Evans placed the perception of grayness in a separate category, that of brilliance [Ref. 2.5, p. 100].

3.5 Brilliance: Grayness and Fluorence

Brilliance, like the attribute lightness, can be perceived only when the object viewed is *not isolated,* for example, an area of paint in a painting or a piece of glass among others in a stained glass window. The perception of brilliance embraces two mutually exclusive aspects: either *grayness* is perceived or what Evans called *fluorence* [Refs. 2.5, p. 99; 3.9], which is an *apparent fluorescence* or *negative grayness* [3.10]. To understand what is implied, let us consider a sheet of paper of a red color that possesses appreciable grayness when it is viewed in a room with normal illumination. In such cases, the light from the surroundings is more intense than that coming from the red sheet to our eyes. If, by means of a spotlight, a continuously increasing amount of light is directed onto the paper while the illumination falling on the surrounding objects remains unchanged, then the grayness of the red paper will decrease progressively and finally reach zero. At this point, the luminance of the light from the paper is still appreciably less than that of the light from the surroundings. This zero point is the separation between the regimes of grayness and fluorence (negative grayness). Then, as the spotlight illumination of the red paper is further increased, fluorence increases from zero (at the zero point) and the red acquires a fluorescent appearance; it is *fluorent* [3.9]. The fluorence continues to increase, but it finally reaches a maximum and then diminishes to zero. The maximum is reached when the lightness of the paper matches that of its surround. Above that lightness, the red appearance of the paper resembles that of a light source [Ref. 2.5, p. 101].

A striking way to experience the grayness aspect of brilliance is to note the grayness of a sheet of neutral gray paper in a well-illuminated room and then the absence of grayness when the lights are turned off and the paper alone is illuminated by a white spotlight. In the latter instance, the paper is viewed in isolation, and the color perceived is white. Similarly, a paper colored brown, which is dark yellow or orange with added grayness, appears yellow or orange when it is viewed in isolation.

It is interesting to note the difference between saturation and brilliance. As mentioned earlier, saturation concerns the relative amount (concentration) of the hue component perceived in a color. Saturation may vary from zero to nearly 100 %. Brilliance, on the other hand, concerns the absolute *amount* of grayness or of negative grayness present, each of which depends on the surroundings.

Evans' recent discovery of brilliance as an attribute of perceived color has thus far received little attention in the current color literature (see, however, [3.4]), yet this attribute should be recognized by artists, designers, and others concerned with the application of color. Evans has pointed to the fact that Pope, in his book *The Language of Drawing and Painting* (1949) [1.22, 3.11], showed an awareness of the need for an attribute such as brilliance [Ref. 2.5, p. 236]. Evans wrote: "There is no question ... of the fundamental soundness of his ideas, nor of the fact that a complete rewriting of that portion of his book in terms of the four variables, hue, saturation, brilliance, and lightness as we have developed them, would remove most, if not all, of the ambiguities he encountered. Carrying out such a work would be a remarkable contribution to the understanding of the arts ..." [Ref. 2.5, p. 235]. According to Evans, when related (nonisolated) colors of nonluminous objects are perceived, only four color attributes are involved: hue, saturation, lightness, and brilliance [Ref. 2.5, p. 137]. Brightness is assigned to the illumination.

As we shall see (Chap. 8), the attributes of perceived colors are sometimes used in important color systems. W.D. Wright [3.12] and A.R. Robertson [3.13] have discussed these attributes as employed in such cases.

3.6 Color Terms

In the science of color perception, the terms "color", "hue", "saturation", "brightness", "lightness", "brilliance", "red", "blue", "achromatic", etc., apply to *color response*. Used in this sense, they are terms of psychology. In the preceding sections, the attributes of color have been described in this sense. In later chapters where color measurement and specification are described, the frame of reference is changed. Color is linked to light (the stimulus) rather than to perception (the response) because precise measurements can be made on light relatively easily. For this reason, a new definition of color has been adopted, *psychophysical color,* that is satisfyingly close to the layman's everyday usage of the term "color". There should be no confusion in this book because the contexts within which the terms are used should provide the necessary clues. Whenever there is a need for clarity or emphasis, however, use is made of the specific terms "psychophysical color" (Sect. 6.1) and "psychological color" [Ref. 3.14, p. 229].

Artists and art writers seem to employ the terms "saturation" and "chroma" interchangeably to denote the purity of a color. The word "Chroma" is from the Munsell color system (Sect. 8.4). It is interesting that Munsell Chroma is intended to be a correlate of perceived saturation, but Evans has shown that it correlates more closely with a combination of saturation and brilliance [Ref. 2.5, p. 168]. In art, the terms "value" ("Value" is used in the Munsell color system) and "tone" are often employed to denote lightness [Ref. 3.15, p. 257].

The term "vividness" applied in art to colors might aptly refer to preceived brilliance. This is also suggested by the word "bright" as in "bright red" [Ref. 2.5, p. 196].

4. Light and Color

4.1 What Is Light?

What is light? A brief answer is: Light is a form of energy. Examples of other forms of energy are kinetic energy, such as that transferred from the wind to the vanes of a windmill, and chemical energy, such as that stored in an automobile battery, available for conversion to electrical energy.

Light is a form of *radiant energy*. More precisely, light is *electromagnetic energy*, a category of radiant energy that includes x-rays, radio waves, etc. In Table 4.1 the various types of radiant energy in the electromagnetic category are presented. The whole range is called the *electromagnetic spectrum*. The

Table 4.1. The electromagnetic spectrum. The visible spectrum occupies a small part of the electromagnetic spectrum. Wavelength is given in kilometers [km], meters [m], centimeters [cm], millimeters [mm], and nanometers [nm]

		-----10 km
		------1 km, 1000 m
	AM	------------100 m
	Short waves	
Radio	TV	-------------10 m
waves	FM	--------------1 m, 100 cm
	Radar	--------------------10 cm
		----------------------1 cm, 10 mm
	Microwaves	--------1 000 000 nm, 0.1 cm, 1 mm
		---------100 000 nm
Infrared radiation		----------10 000 nm
		------------1 000 nm
Visible radiation		--------------100 nm
Ultraviolet radiation		--------------10 nm
		--------------1 nm
X-rays		---------------0.1 nm
		--------------0.01 nm
Gamma rays		---------------0.001 nm

relatively small range within it that represents *visible* radiant energy, is called the *visible spectrum*. We commonly define light as *visible radiant energy* [Note 4.1].

The term *visible radiant energy* for light implies correctly that the visual system responds to it in the experience of seeing. We know that it does not respond to radio waves. Nor does it respond to infrared radiation, ultraviolet radiation[1], x-rays, and gamma rays, but eyesight can be destroyed by them. Only light is the *stimulus* to vision.

The portion of the sun's radiation that penetrates the earth's atmosphere consists principally of visible, infrared, and ultraviolet radiation. This "mixture" reaches the earth's surface not only directly as sunbeams but also indirectly by scattering from water droplets in clouds and from dust particles, and by scattering produced by molecules (mainly nitrogen and oxygen) in the atmosphere (Sect. 2.2). As a result, infrared, ultraviolet, and visible radiation in various proportions falls on the earth from blue, hazy, and overcast skies. The radiation emitted by the hot tungsten filament of a common light bulb (incandescent lamp) and by the phosphors of a fluorescent lamp contains not only visible and infrared radiation but also some ultraviolet radiation.

4.2 Wavelength and Light

Physicists tell us that electromagnetic radiation possesses a wavelike character. Indeed, measures of waves, such as *wavelength* and *wave frequency*, are used in the measurement of electromagnetic radiation. Only wavelength is used in discussions in this book, because it is the measure most commonly found in the literature on color. The classifications in Table 4.1 have been made on the basis of wavelength.

Those who are familiar with the operation of radios know that if the wavelength of radio waves is reported, it is given in meters and kilometers. In the case of light, for which wavelengths are very much shorter, the unit of length commonly used is the *nanometer* nm. One nanometer is equal to one millionth of a millimeter (a millimeter is one tenth of a centimeter) and to one billionth (USA) or one thousand millionth (UK) of a meter. Until recently, in the literature on color, the use of units called millimicrons and angstroms was common. One nanometer equals one millimicron; one nanometer equals ten angstroms.

---

[1] A portion of the spectrum of ultraviolet radiation is sometimes called "black light" because it is invisible ("black") in a darkened room and yet it excites fluorescence in many materials, causing them to glow and be visible in the dark (Sect. 5.4). It also affects photographic film in the dark.

4.3 Spectral and Nonspectral Hues

Visible radiation is commonly considered to be represented in the electromagnetic spectrum in the wavelength range between 380 and 780 nm (Tables 4.1,2). A significant question is: What is perceived when light of a single wavelength (say 500 nm) is viewed? The answer is: Green. At 600 nm, it is reddish orange; at 470 nm, blue. Table 4.2 shows the hues perceived for radiation over the whole visible range. Actually the hues change gradually when the wavelength is increased continuously from 380 to 780 nm. Thus, the greenish blue at 486 nm is more greenish than the greenish blue at 483 nm. Light of a single wavelength is called *monochromatic light*.

The colors of the spectrum produced by monochromatic light have maximum saturation. Some investigators in color science consider that these colors contain an achromatic component and hence that they do not possess 100 % saturation (Sect. 11.3) [4.2,3]. Evans, on the other hand, believed that

Table 4.2. The visible spectrum and the nonspectral range

	Color names for lights*	Hue wavelength range [nm]	Hue complementary wavelength range‡ [nm]	Relative luminosity (Sect. 4.7)
	Bluish purple (bP) (violet)†	380	563c	0.0001
	Purplish blue(pB)(blue violet)†	-430	------	0.0116
	Blue (B)	-465	------	0.075
	Greenish blue (gB)	-482	------	0.15
	Blue green (BG)	-487	------	0.18
Spectral colors (visible spectrum)	Bluish green (bG)	-493	------	0.24
	Green (G)	-498	------	0.29
	Yellowish green (yG)	-530	------	0.862
	Yellow green (YG)	-558	------	1.00
	Greenish yellow (gY)	-570	------	0.952
	Yellow (Y)	-575	------	0.91
	Yellowish orange (yO)	-580	------	0.87
	Orange (O)	-586	------	0.80
	Reddish orange (rO)	-596	------	0.68
	Red (R)	-620	------	0.381
	Red (R)	680	492c	0.17
	Red (R)		494c	
	Purplish red (pR)		498c	
Nonspectral colors	Red purple (RP)		528c	
	Reddish purple (rP)		553c	
	Purple (P)		563c	
	Bluish purple (bP)			

* Names for colored lights proposed by Kelly [4.1].
† Names employed by other authors [4.1].
‡ Complementary wavelength with respect to CIE ILL C.(Sect. 6.4)

the colors produced by monochromatic light do not contain an achromatic component, with the possible exception of yellow [Ref. 2.5, pp. 73, 121].

Most light that we experience is not monochromatic. For example, a beam of blue light from a colored lamp may be found to contain light of wavelengths ranging over half of the visible spectrum. The major difference between a green beam and a blue beam from ordinary colored lamps is in the relative amounts of light contained in the green and blue wavelength regions. For example, a green beam has typically relatively larger amounts of light in the green region, from 500 to 550 nm, and a blue beam has larger amounts in the blue region, from 400 to 500 nm. The particular hue that is perceived is due partly to the predominance of energy in a wavelength region and partly to the brightness sensitivity of the eye (Sect. 4.7). Much of the light present at other wavelengths can be considered to "cancel out" chromatically by the mixture of additive complementary wavelengths (Sect. 7.2) producing an achromatic component that dilutes the dominant hue and hence lowers the saturation.

When a beam of light from a lamp or from the sun is passed into an optical device called a *monochromator,* portions of the radiation can be isolated in *wavelength intervals* or *bands,* for example, in 10-nm wavelength intervals. Thus, a wavelength interval of light of 500–510 nm can be separated and projected onto a screen. In a device called a *spectroradiometer* [Refs. 1.20, p. 148; 2.1, p. 11; 4.4, p. 29], each wavelength interval of light over the visible range from 380 to 780 nm is separated and the rate at which radiant energy is received (*power*) within each interval is measured. A graph of power versus wavelength for the green and blue lamps discussed above provides a comparison of their *wavelength compositions.* Usually, however, the power data are converted to relative values to facilitate comparisons of their chromatic qualities, especially when the magnitudes of the powers of the light sources are appreciably different, e.g., a 25-watt (W) lamp, a 1000 W lamp, sunlight, etc. (Sect. 4.5).

The hues represented by monochromatic radiation frqm 380 to 780 nm are those that are present in the sun's visible spectrum, a common example of which is provided by a rainbow. Those hues are called the *spectral hues;* all colors, regardless of saturation (the colors in a rainbow have low saturation), that are perceived to have a spectral hue are called *spectral colors.* But spectral hues are not the only ones that we commonly experience. There are also purple, purplish red, and a range of neighboring red hues that are not present in the sun's spectrum or in the spectrum of any source. Such hues are called *nonspectral hues;* colors having these hues are called *nonspectral colors.* Monochromatic radiation cannot produce nonspectral colors, but mixtures of two or more beams of monochromatic radiation of different wavelength can. Nonspectral colors of maximum saturation can be produced by combinations of, for example, monochromatic light of wavelength 680 nm (red) and monochromatic light of wavelength 420 nm (bluish purple). Nonspectral colors of lower saturation are produced by beams that commonly contain light from most of the spectral range but with predominant amounts from the red and blue regions. A comprehension of the wavelength composition of light is aided very much by graphical presentations, which are discussed in Sect. 4.5.

4.4 Light from Lasers

Relatively recently, light sources called *lasers* have become commercially available. They can be used to produce monochromatic light. Because their light beams (called *laser beams*) commonly have radiation densities that greatly exceed those of light beams from ordinary lamps and of sunbeams, lasers are finding diverse uses in science, medicine, and technology. A laser beam consists of light in parallel rays and of one, two, or several wavelengths. The light is said to be coherent, which means that wave trains of energy are in step with each other, not out of step (out of phase) as in ordinary light [4.5].

Various media are used for the production of laser beams: crystals, glasses, gases (for example, argon, krypton, and mixtures of helium and neon), and solutions of dyes. Some *gas lasers* have been used as light sources in light art. Frequently, a helium-neon laser is employed, which produces a beam of monochromatic red light (632.8 nm) [4.6,7]. Also, the use of an argon gas laser with a beam containing principally two wavelengths (488.0 nm, blue green, and 514.5 nm, green) has been reported [4.6]. In the latter case, light of the two wavelengths was separated by a diffraction grating to produce two monochromatic beams of different color [4.8].

Dye lasers can produce beams at any desired wavelength between 400 and 750 nm [4.9–12]. At their present stage of development, dye lasers are exclusively pulsed devices; they require auxiliary equipment (an electronic flash tube or an additional laser) to drive them [4.13]. Of particular interest is the tunability of dye lasers, which permits the production of monochromatic light of any wavelength within ranges of 30 to 50 nm or more [4.11,12].

In the selection of lasers, very careful consideration must be given to their safety hazards. Outputs from lasers at power levels below 5 milliwatts (mW) have been employed in art displays, but even at such levels, certain precautions must be taken to ensure acceptable safety [Refs. 4.14, p. 7; 4.15, p. 100].

4.5 Light from the Sun and from Lamps

As mentioned in Sect. 4.3, most light that we experience is not monochromatic; an example of typical green and blue lights was cited. It is characteristic of various light sources (the sun, a candle flame, a light bulb with an incandescent tungsten filament, a fluorescent lamp, etc.) that there are appreciable differences in the rates at which radiant energy is emitted (power) in wavelength intervals (say, of 10 nm) over the range from 380 to 780 nm. As noted before in Sect. 4.3, the wavelength composition of the radiation emitted by a source is most conveniently displayed by graphs showing *relative power* versus wavelength nm (Figs. 4.1,2). Measurements of power are converted to relative power based in general on the convention that relative power has a fixed value of 100 at 560-nm wavelength. Thus, curves showing relative power for most light sources intersect at 560 nm [Note 4.2].

The wavelength-composition graph showing relative power plotted versus wavelength is called a *relative spectral power distribution curve*. (In earlier publications on illumination and color, the term "relative energy" was used; it has been replaced by "relative power".) Typical curves for light from an incandescent-tungsten-filament lamp and from a fluorescent lamp are shown in Figs. 4.1,2. Comparison of the two curves reveals the relatively greater amount of radiation at 450 nm for the fluorescent lamp and at 650 nm for the incandescent light bulb. From the shapes of the two curves near 380 nm, it is clear that both extend to wavelengths below 380 nm and,

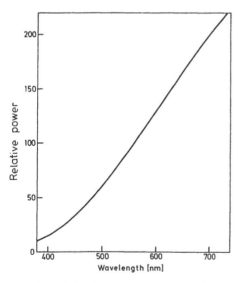

Fig. 4.1. Wavelength composition of light from a tungsten-filament lamp [typified by CIE ILL A (Sect. 4.6)]. Relative spectral power distribution curve. Color temperature: 2856 K

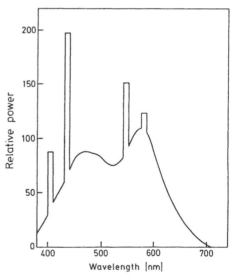

Fig. 4.2. Wavelength composition of light from a daylight fluorescent lamp. Typical relative spectral power distribution curve. Correlated color temperature: 6000 K. (Based on data of Jerome reported in [Ref. 3.14, p. 37])

hence, that the radiation from such fluorescent and incandescent lamps includes ultraviolet radiation.

The relative spectral power distribution curve for the daylight-fluorescent-lamp radiation shows four vertical bars (Fig. 4.2). Each represents a wavelength interval, 10 nm wide, within which there is a tall sharp peak or jump of radiation that is characteristic for mercury vapor, which is in the tube. (The peaks occur at wavelengths of approximately 405, 436, 546, and 578 nm [4.16].) The smooth, continuous portions of the curve represent the radiation contributed by the phosphors in the lamp. The jumps, four monochromatic emissions from the mercury, are superimposed on, or mixed with, the diffuse multicomponent contribution from the phosphors. A precise indication of the relative magnitudes of the actual peaks would serve no useful purpose in the discussion of color. It is sufficient that each bar shown represents accurately the power averaged over a 10-nm wavelength interval.

Figure 4.3 shows typical relative spectral power distribution curves for direct sunlight (I) and for north-sky light received on a 45° plane (II) at Cleveland, Ohio [4.17]. These two curves may be compared with a standard curve CIE ILL D$_{65}$ (Sect. 4.6), which represents a typical phase of daylight. North-sky light in the northern hemisphere is judged to be "cooler" than direct sunlight, because it contains a greater proportion of light at shorter wavelengths (blue) and a lower proportion of light at longer wavelengths (red). Also shown in Fig. 4.3 is a horizontal dashed line (E) which has been added to represent an *equal power distribution* – that is, a distribution in which the relative power does not vary with wavelength. This distribution

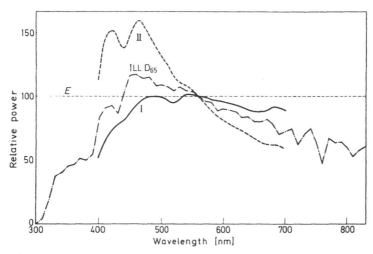

Fig. 4.3. Wavelength composition of direct sunlight (*I*) and north-sky light (*II*). The curves for CIE ILL D$_{65}$ (Sect. 4.6) and the equal power distribution (*E*) are shown for comparison. Relative spectral power distribution curves. (Curves *I* and *II* are based on observations in Cleveland, Ohio, reported in [4.17])

serves as an arbitrary definition of a white light for purposes discussed later (Sects. 6.3, 7.2). Generally it is of interest because it can be regarded as a kind of intermediate representation for white light between the extremes of daylight and ordinary incandescent-lamp illumination [Ref. 2.5, p. 52]. A hypothetical light source capable of producing an equal power distribution (*equal-energy light*) is usually called an *equal-energy light source.*

4.6 Standard Illuminants (CIE)

Because the perceived colors of objects generally vary with the illumination in which they are viewed, we tend to prefer to make color comparisons in daylight. But in color specification and color measurement, the wavelength composition of daylight must be specified precisely. For this reason, it has been found practical to establish internationally acceptable standards in the form of arbitrary wavelength compositions that represent typical sunlight, daylight, and artificial illumination.

These standards, called *CIE Illuminants,* have been established by the Commission Internationale d'Eclairage (CIE). (In the 1930s and 1940s, it was common in the United States to refer to the Commission by its English name or initials, International Commission on Illumination, I.C.I., but they are no longer used [Ref. 4.18, p. 4].) It is to be emphasized that the standard illuminants are, in reality, tables of numbers that state fixed wavelength compositions. Light having some of these wavelength compositions is produced in color-measurement laboratories with the use of special lamps and filters. Figures 4.4,5 show plots that represent several important CIE Illuminants.

One illuminant, called CIE Illuminant A, or simply *CIE ILL A,* represents closely the wavelength composition of light from a 500-W tungsten-filament light bulb (2856 K, Table 7.6) [Ref. 3.14, p. 47]. The relative spectral power distribution curve for CIE ILL A is given in Figs. 4.1,4. Another illuminant, *CIE ILL B,* typifies the wavelength composition of direct sunlight at noon. The illuminant *CIE ILL C* is particularly important, because its wavelength composition is a close approximation to that of average daylight [Ref. 4.4, p. 7]. Most color measurements from the 1930s to the 1960s were reported in terms of CIE ILL C; some useful tools for considering colors relate to this illuminant (Sects. 7.1,5; 8.4–6,8; 10.1).

The wavelength compositions of sunlight and of daylight are represented by CIE ILL B and CIE ILL C rather well, but only in the range from 400 to 700 nm. For the color measurement of fluorescent materials, illuminants should be used whose relative spectral power distributions in the wavelength range from 300 to 400 nm also typify those of sunlight and daylight. A new series of standard illuminants was introduced more recently that

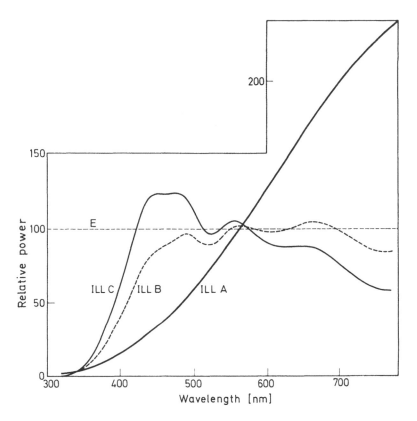

Fig. 4.4. CIE Illuminants (CIE ILL A, CIE ILL B, and CIE ILL C) and the equal power distribution (E). Relative spectral power distribution curves

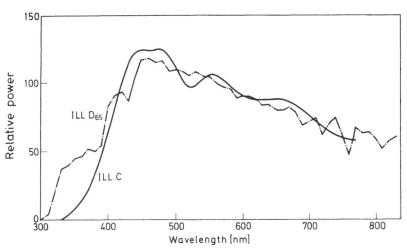

Fig. 4.5. CIE Illuminants (CIE ILL D_{65} and CIE ILL C). Relative spectral power distribution curves [4.19]

represent well (from 300 to 830 nm) the wavelength compositions of various phases of daylight, the most common being CIE ILL D_{55}, CIE ILL D_{65}, and CIE ILL D_{75} [4.19]. [The subscripts 55, 65, and 75 are keys referring to the color temperatures of the illuminants; 5503, 6504, and 7504 K, respectively (Table 7.6).]

For the most part, CIE ILL C has been replaced by *CIE ILL D_{65}*, which represents an average phase of daylight over the expanded wavelength range from 300 to 830 nm [Ref. 1.18, p. 110]. In Fig. 4.5, the relative spectral power distribution curves for CIE ILL C and CIE ILL D_{65} [4.19] can be compared. It is clear that only in the ultraviolet region, below 380 nm, do the two curves differ significantly.

It is useful to know about the principal standard illuminants, because they are commonly part of technical color specifications. In a color specification, the illuminant describes the illumination of an object for which the stated color applies.

4.7 Eye Brightness Sensitivity

Light is described as visible radiation [Note 4.1]. In Sect. 4.3 it was pointed out that the visible range of radiation extends from 380 to 780 nm. In general, a normal eye is essentially blind to all radiation of wavelengths shorter than 380 nm and longer than 780 nm. How well does the normal eye respond within the range of visibility?

Thorough investigations have been performed to provide the answer. They show that the brightness sensitivity of a normal eye to monochromatic light increases as the wavelength is increased, starting from zero sensation at about 380 nm. The brightness sensitivity reaches a maximum at about 555 nm and then decreases, reaching zero at about 780 nm. The internationally accepted set of data representing an "average" normal eye adapted to daylight is presented in part in Table 4.2. These data are referred to as *relative luminosities*. A graph showing brightness sensitivity versus wavelength is a bell-shaped curve that has its maximum (1.000) at a wavelength of 555 nm. It is curve (II) in Fig. 6.8 and the black curve in Plate XIV; it is used in color-measurement calculations (Sect. 6.3).

The bell-shaped curve shows that at wavelengths of 510 and 610 nm the relative luminosity is 0.500. This implies that a typical normal eye is half as sensitive to radiation at wavelengths 510 and 610 nm as it is to radiation at 555 nm. The hue response to monochromatic radiation at 555 nm is yellowish green; at 510 nm, green; and at 610 nm, reddish orange. At 472 nm (blue) and at 650 nm (red) the brightness sensitivities of the normal human eye are about one-tenth its brightness sensitivity at 555 nm.

The eye brightness sensitivity discussed briefly above applies to situations in which the average normal eye is adapted to a high level of illumination, such as that encountered in daylight or in the usual artificial illumination sufficient, say, for reading. At high illumination levels, the cones of the retina function (photopic vision), enabling color vision (Sect. 2.2). At very low illumination levels, when the rods of the retina determine vision (scotopic vision), the eye brightness sensitivity to visible radiant energy is significantly different. Again, a bell-shaped curve applies, but it is located in a shorter wavelength range; the maxiumum eye brightness sensitivity occurs at 507 nm [Ref. 1.20, p. 60]. Because the topics in this book concern only color and color vision, I shall refer only to the eye brightness sensitivity for the average normal eye adapted to a high level of illumination (photopic vision).

5. Colored Materials

5.1 Pigments and Dyes

Substances that are added to materials to produce colors (pigments and dyes) are called *colorants*. *Dyes* are soluble substances, or substances that are soluble during a stage of a dyeing process. They are commonly added to textiles, paper, plastics, leather, etc. Certain classes of dyes are rendered insoluble by a chemical process after they have penetrated into the material being dyed. *Pigments* are insoluble substances, particles of which are dispersed in paints, lacquers, inks, paper, plastics, rubber, etc. *Fluorescent pigments* for paints are commonly fluorescent dyes dissolved in a solid plastic. (The dye and plastic ingredients are combined before the plastic is transformed chemically into an insoluble solid and ground to a pigment powder.)

Thousands of dyes and pigments are in current use. Of these, a relatively small number are of interest to artists and designers. H.W. Levison, in his book on the lightfastness of artists' pigments, lists about 100 pigments of which a few are now being used by artists, most of the others being of potential use to them [5.1]. Information about artists' pigments, both past and present, may be found in handbooks by Mayer [5.2], Wehlte [5.3], and Gettens and Stout [5.4] and in a chapter by Levison [5.5].

A comprehensive listing of virtually all (8 000) current industrial colorants is provided in the *Colour Index* (C.I.) (3rd edition) in five volumes; a sixth volume contains supplementary information [5.6,7]. Volumes I, II, and III list all colorants according to their recognized usages. Each colorant is given a C.I. Name (generic) and C.I. Number, which together code its usage (e.g., as an acid dye, solvent dye, pigment), and a C.I. Constitution Number which codes its chemical classification. (For example, for a phthalocyanine blue the C.I. Name and Number are C.I. Pigment Blue 16, and the C.I. Constitution Number is 74 100.) Appropriate information (hue, fastness, solubility, etc.) is tabulated for the various usages. In Volume IV, technical information is given on the chemical constitution and preparation of the colorants, and literature sources are listed. Volume V consists of a list of manufacturers, a C.I. Names (generic) Index and a Commercial Names (Trade Names) Index. Because new colorants frequently become commercially available and lists of manufacturers and trade names require periodic revision, the *Colour Index* is kept up-to-date by the publication *Additions and Amendments*.

5.2 Opaque Materials

Most objects that we view are opaque. Light that falls on a nonfluorescent opaque paint film is affected by the paint's pigments and vehicle in three different ways. Part of the incident light is reflected away without entering the surface (*surface reflection*). The wavelength composition of surface-reflected light is virtually unchanged by the paint; it is nearly identical to that of the incident light. If the paint surface is *matt, diffuse surface reflection* occurs; the rough pigment particles that poke out of the paint surface cause surface reflection in all directions. If the surface is smooth and glossy, then the surface reflection of incident light is mirrorlike, which is called *specular reflection* [Ref. 1.20, p. 36]. However, diffuse ambient light that arrives from all directions at a glossy surface is, of course, reflected in all directions.

The rest of the light, usually the major part, penetrates the surface. The light passes through pigment particles in which some is *selectively absorbed* while the remainder is *scattered* diffusely to the surroundings. The absorbed light is converted to heat and disappears unnoticed. Selective light absorption by pigments and dyes implies that light is absorbed in varying degrees depending on wavelength. The wavelength dependency of the absorption is determined by chemistry, by the specific molecular architecture that characterizes an individual colorant. As a consequence, the wavelength composition of the unabsorbed light that leaves a pigment particle differs from that of the incident light and from that of the unabsorbed light that leaves particles of other pigments.

If a paint were made that contained a white pigment of perfect whiteness, light that penetrated the pigment particles would not be absorbed. The light would be scattered diffusely from within the paint film and the scattered light would have the same wavelength composition as the incident light. High-quality white pigments do absorb light selectively, but only in a minor way.

Thus, the three processes affecting light incident on an opaque (nonfluorescent) paint film are surface reflection, selective absorption, and scattering. If daylight (white light) falls on a matt paint film pigmented with cadmium red, the light that finally reaches our eyes is a mixture of white light (surface-reflected light) and red light (light that is scattered after selective absorption has occurred). The perceived color, also called *object color*, or more specifically, *surface color*, is red. The purity of the red depends on both the character of the selective absorption and the extent of dilution by the surface-reflected light.

When light falls on an opaque surface, the fractional amount (or percentage) that is reflected and scattered (the remainder being lost by absorption) is called the *reflectance* [Ref. 2.1, p. 187]. Reflectance is commonly measured with a *spectrophotometer* [Refs. 1.39, p. 81; 4.4, pp. 3, 37; 5.8, p. 74]. Such measurements are usually made at 5- or 10-nm intervals over the whole visible range (380–780 nm), and the data are presented graphically as *spectral reflectance curves* (Figs. 5.1–9). Each curve is a quantitative descriptive record of an individual sample (paint film, pigment powder, dyed plastic, etc.). It is usual to find that, although different samples of pigment powder (say, cadmium red) may be chemically identical, slight variations occur in their spectral reflectance curves. Such variations may be introduced not only by the presence of impurities but also by insufficient crystal growth and crystal distortions resulting from processing. Spectral reflectance curves are essential tools in an important method of color measurement (Sects. 6.1,3). They are also useful in estimating wavelength compositions of light in judging the results of color mixing (Sect. 5.7).

The spectral reflectance curves shown in Figs. 5.1–9 refer to films containing pigments, most of which are among those commonly used by artists. In one series of tests made by N.F. Barnes (Figs. 5.1–4), only a minimum amount of vehicle (a glue) was used to hold the pigment particles together to form a matt pigment film [5.9,10]. The CIE color designations of the films are given in Table 5.1. In another investigation by H.R. Davidson (Figs. 5.5–9), glossy films were employed with an acrylic resin serving as vehicle [5.11].

Table 5.1. CIE color designations of matt pigment films (CIE ILL C) represented in Figs. 5.1–4. Data of Barnes [5.9,10]. See Sects. 6.3,4 for explanation of x, y, Y, λ_D, λ_c, and p_e

Pigment used in film	Abbreviation	x	y	Y [%]	λ_D, λ_c [nm]	p_e [%]
Cadmium red	CR	0.5375	0.3402	20.78	604.8	67.3
Madder lake	ML	0.3985	0.2756	33.55	496.5c	32.9
Cadmium orange medium	COM	0.5245	0.4260	42.18	586.9	86.9
Cadmium yellow light	CYL	0.4500	0.4819	76.66	575.3	81.9
Zinc yellow	ZY	0.4486	0.4746	82.57	575.8	79.7
Yellow ochre	YO	0.4303	0.4045	41.15	581.9	55.9
Emerald green	EG	0.2446	0.4215	39.12	511.9	22.8
Terre verte	TV	0.3092	0.3510	29.04	549.8	9.2
Viridian	V	0.2167	0.3635	9.85	497.1	31.9
Cobalt blue	CB	0.1798	0.1641	16.81	474.6	65.5
Ultramarine blue (natural)	UB	0.2126	0.2016	18.64	474.4	49.2
French ultramarine blue (artificial)	FUB	0.1747	0.1151	7.84	467.8	75.1
Manganese violet	MV	0.3073	0.2612	26.99	553.7c	21.6

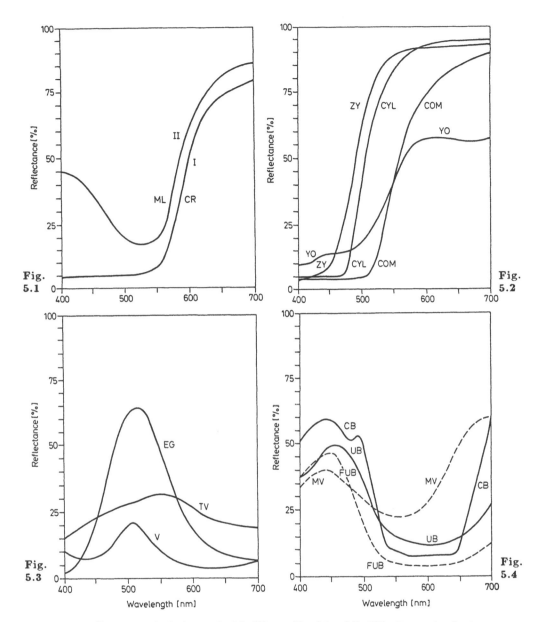

Fig. 5.1. Pigments. Cadmium red, CR (*I*); madder lake, ML (*II*). Spectral reflectance curves for matt pigment films. The CIE color designations are given in Table 5.1. (Based on curves in [5.9,10])

Fig. 5.2. Pigments. Cadmium orange medium, COM; cadmium yellow light, CYL; zinc yellow, ZY; yellow ochre, YO. Spectral reflectance curves for matt pigment films. The CIE color designations are given in Table 5.1. (Based on curves in [5.9,10])

Fig. 5.3. Pigments. Emerald green, EG; terre verte, TV; viridian, V. Spectral reflectance curves for matt pigment films. The CIE color designations are given in Table 5.1. (Based on curves in [5.9,10])

Fig. 5.4. Pigments. Cobalt blue, CB; ultramarine blue (natural), UB; French ultramarine blue (artificial), FUB; manganese violet, MV. Spectral reflectance curves for matt pigment films. The CIE color designations are given in Table 5.1. (Based on curves in [5.9,10])

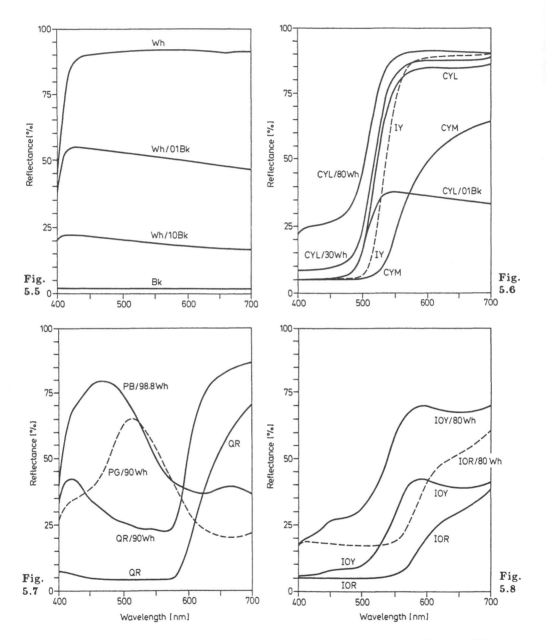

Fig. 5.5. Pigments and pigment mixtures. Carbon black, Bk; titanium dioxide white, Wh; mixture Wh/01Bk (99 % Wh/1 % Bk); mixture Wh/10Bk (90 % Wh/10 % Bk). Spectral reflectance curves for glossy paint films (OSA-UCS) (Chap. 9). (Based on data in [5.11])

Fig. 5.6. Pigments and pigment mixtures. Chrome yellow light, CYL; chrome yellow medium, CYM; indolinone yellow, IY; carbon black, Bk (Fig. 5.5); titanium dioxide white, Wh (Fig. 5.5); mixture CYL/30Wh (70 % CYL/30 % Wh); mixture CYL/80W (20 %CYL/80 % Wh); mixture CYL/01Bk (99 % CYL/1 % Bk). Spectral reflectance curves for glossy paint films (OSA-UCS) (Chap. 9). (Based on data in [5.11])

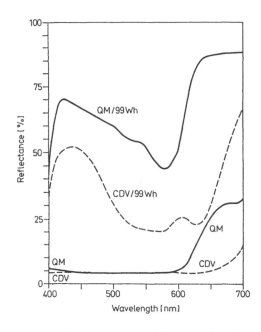

Fig. 5.9. Pigments and pigment mixtures. Quinacridone magenta, QM; carbazole dioxazine violet, CDV; titanium dioxide white, Wh (Fig. 5.5); mixture QM/99Wh (1 % QM/99 % Wh); mixture CDV/99Wh (1 % CDV/99 % Wh). Spectral reflectance curves for glossy paint films (OSA-UCS) (Chap. 9). (Based on data in [5.11])

For purposes of illustration, let us consider first the curve (I) for cadmium red in Fig. 5.1. It shows that the pigment exposed to daylight absorbs less than half of the light that it receives in the wavelength range above 600 nm and absorbs most of the light at wavelengths below 600 nm. Hence, mostly light of longer wavelengths (red) is scattered to the eye. Madder lake (II), however, absorbs smaller proportions of the light at wavelengths both above 600 nm (red) and below 480 nm (blue). The result is that when madder lake is illuminated by daylight a mixture of mostly long- and short-wavelength light (magenta) passes from the pigment particles to the eye.

The spectral reflectance curves in Figs. 5.1–9 represent the simultaneous occurrence of diffuse surface reflection and selective absorption at wavelengths from 400 to 700 nm. Thus the scattered light that comes from selective absorption within the pigment particles is diluted at each wavelength by some surface-reflected daylight (typified for example by CIE ILL C). The spectral reflectance curve in the case of the cadmium red pigment (I) shows that below 550 nm about 5 % of the incident light was not absorbed. It

Fig. 5.7. Pigments and pigment mixtures. Phthalocyanine blue, PB; phthalocyanine green, PG; quinacridone red, QR; titanium dioxide white, Wh (Fig. 5.5); mixture PB/98.8Wh (1.2 % PB/98.8 % Wh); mixture PG/90Wh (10 % PG/90 % Wh); mixture QR/90Wh (10 % QR/90 % Wh). Over the range 400–700 nm, the reflectances of PB and PG vary between 4.1 and 6.5 %. Spectral reflectance curves for glossy paint films (OSA-UCS) (Chap. 9). (Based on data in [5.11])

Fig. 5.8. Pigments and pigment mixtures. Iron oxide yellow, IOY; iron oxide red, IOR; titanium dioxide white, Wh (Fig. 5.5); mixture IOY/80Wh (20 % IOY/80 % Wh); mixture IOR/80Wh (20 % IOR/80 % Wh). Spectral reflectance curves for glossy paint films (OSA-UCS) (Chap. 9). (Based on data in [5.11])

33

is possible that surface-reflected daylight accounted for most of the 5%. Davidson states that his data for high-gloss samples may be converted to data in which surface reflection is excluded by subtracting about 4% from the reflectance at each wavelength [5.11].

Glossy paint films and varnished oil paintings often have colors of greater saturation than those of matt films that contain the same pigment. The reason is that, in the former, the pigment particles at the paint film surface are covered by a smooth glossy layer of paint vehicle (for example, dried linseed oil) or of varnish resin. In a red glossy paint film, for example, some of the scattered red light is reflected from the smooth surface back into the pigment particles again (*internal reflection*) where it is subjected to further selective absorption. This red light, which results from two passages of the light through the film, is combined with the red light that is not internally reflected and with surface-reflected white light to produce a mixture of light whose color is more saturated than the color obtainable with the same pigment in a matt film [Ref. 2.1, p. 283].

The saturation of the color can be increased by viewing the glossy film when it is illuminated by a direct beam of light, such as a sunbeam. Even better results can be obtained by illumination with a projector beam in a darkened room, to avoid illumination with diffuse ambient light, which could produce diffuse, desaturating reflection in all directions. With isolated direct lighting, the white light that is specularly reflected from the surface, as from a mirror, can easily be avoided by a viewer, so that only undiluted scattered red light reaches the eye [Ref. 2.1, p. 282].

The spectral reflectance curves in Figs. 5.1–9 are not accurate indicators of the wavelength composition of the light that comes from illuminated cadmium red and madder lake pigments [Note 5.1]. The wavelength composition of the light that leaves a pigment film depends not only on the

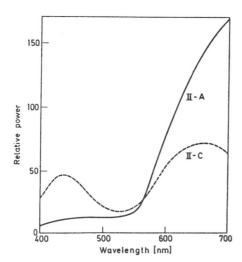

Fig. 5.10. Madder lake pigment. Wavelength composition of light reflected from a matt pigment film (madder lake pigment [5.9,10]) when illuminated by daylight (*II-C*), typified by CIE ILL C, and by incandescent-lamp light (*II-A*), typified by CIE ILL A. Relative spectral power distribution curves. See Fig. 5.1

absorption characteristics of the pigment and on the surface reflection but also on the wavelength composition of the incident light. The effect of the incident light on the wavelength composition of the light from a pigment film is demonstrated in Fig. 5.10. Curve II-C is the relative spectral power distribution curve for the light coming from a madder lake film when it is exposed to daylight (typified by CIE ILL C). Curve II-A is for light that comes from the same pigment film when it is illuminated by an incandescent-tungsten-filament lamp (typified by CIE ILL A). The marked difference in the wavelength compositions of CIE ILL C and CIE ILL A can be seen in Fig. 4.4. Because light from incandescent lamps has relatively high radiant power at longer wavelengths (reds) and relatively low radiant power at shorter wavelengths (blues) (such light is often said to be "warm"), the light scattered by the pigment is richer in longer wavelengths and poorer in shorter wavelengths. Thus, in incandescent-lamp illumination, the madder lake pigment appears red, not magenta, because, as curve II-A shows, the blue content of the scattered light is very low.

5.3 Transparent Materials

The phenomena that occur when light falls on a *transparent* material, such as colored glass or plastic, are essentially the same as those that occur when light falls on an opaque paint film. Part of the light that passes through the transparent (nonfluorescent) material is absorbed and dissipated, unnoticed, as heat; the remainder that is not absorbed emerges from the opposite side as *transmitted light*. (I am not now taking into account the simultaneous occurrence of internal reflection.) When daylight enters one side of a colored glass and red light is transmitted from the other side to our eyes, we say that the glass is red – that is, the perceived color (object color) is red.

There is one difference to be noted, however. If an opaque matt red paint film is viewed in daylight, the red light that reaches our eyes from within the pigment particles is diluted by daylight that is reflected diffusely from the surface of the film. In the case of a transparent red glass, for example, there is also surface reflection of diffuse daylight, but this white light does not mix with the transmitted red light because it reflects from the glass in the opposite direction – that is, away from our eyes.

A *spectral transmittance curve* for a red purple glass of 1-mm thickness is shown in Fig. 5.11. The *transmittance* of a transparent material is the fraction or percentage of the incident light that passes completely through the material. A spectral transmittance curve is analogous to a spectral reflectance curve for an opaque material. The red purple glass allows red light (wavelengths longer than 630 nm) and blue light (wavelengths shorter than 480 nm) to pass through it; like the pigment madder lake, it absorbs most of the light of intermediate wavelengths.

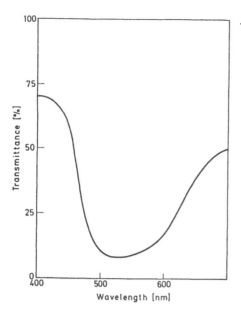

Transmittance [%]

Wavelength [nm]

◀ **Fig. 5.11.** Red purple glass. 1-mm thickness. Spectral transmittance curve

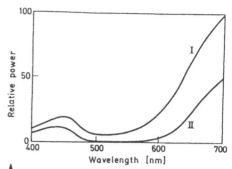

Relative power

Wavelength [nm]

Fig. 5.12. Red purple glass. Wavelength composition of light after it passes through 1-mm (*I*) and 2-mm (*II*) thick glass. The light is from an incandescent lamp and is typified by CIE ILL A. Relative spectral power distribution curves

Figure 5.12 shows the relative spectral power distribution (curve I) for the light that emerges from the red purple glass of 1-mm thickness when it is illuminated by an incandescent-tungsten-filament lamp (typified by CIE ILL A) [Note 5.1]. Because the relative amount of blue light (radiant power at lower wavelengths) in the lamp light is low (Fig. 4.1), only a small amount of blue light is transmitted, as the small hump at the left end of curve I indicates. As a result, the color produced is reddish (pink) (chromaticity: 0.440, 0.279) (Fig. 7.1). [When daylight (CIE ILL C) passes through the glass, the color observed is purple (chromaticity: 0.261, 0.144).] Curve II shows the relative spectral power distribution of the light that is transmitted in the case of two layers of glass or of one layer 2-mm thick. It is significant that the curves (Fig. 5.12) move to lower positions when the number of layers or the layer thickness increases: the brightness of the transmitted light diminishes. It is also significant that the "valleys" decrease to lower positions faster than the "summits" (of the humps) [Note 5.2]; as a consequence the saturation increases and the hue may change.

Now we might ask what occurs in the case of a transparent colored paint film on white paper. Let us consider the case of a paint film that has the characteristics shown in Fig. 5.11. If light from an incandescent-tungsten-filament lamp falls on the paint film, curve I in Fig. 5.12 could represent the wavelength composition of the pink light that reaches the surface of the paper after traversing the film. If we can assume that the white paper reflects all of the light that reaches it and that the reflection is diffuse, then the light will pass through the paint film again, but in the opposite direction, and will emerge as red purple light (chromaticity: 0.383, 0.168) having the wavelength composition given by curve II (Fig. 5.12) on reaching the film surface. The red purple light that emerges from the film surface will mix with the diffuse surface-reflected light (say 4 % of the incident light), and then, instead of curve II, another curve that includes the contribution of surface-reflected light will represent the total reflected light.

5.4 Fluorescent Materials

Paints, inks, and plastics that contain fluorescent dyes are used commonly in advertising and decoration. The California artist Richard Bowman began using fluorescent lacquers in his paintings in 1950 [5.12]; the use of fluorescent paints and inks in art has been spreading widely. Many of the colors produced by fluorescent dyes viewed in, say, daylight cannot be produced by nonfluorescent dyes and pigments under the same conditions (Sect. 7.6).

In what way do the phenomena that occur in nonfluorescent and fluorescent materials differ when they are illuminated by daylight or lamplight? Here it is helpful to recall that, after light penetrates into an opaque or transparent *nonfluorescent* material, part of the received light is absorbed selectively and the remainder is scattered back or transmitted. The absorbed light is transformed to heat, which disappears unnoticed. It should be added that ultraviolet radiation that passes into a nonfluorescent material undergoes the same changes: part is absorbed selectively and transformed completely to heat, and the remainder is scattered back or transmitted as invisible radiation (ultraviolet).

When visible radiant energy (light) and ultraviolet radiation penetrate into a *fluorescent* material, again some is scattered back or transmitted and the remainder is absorbed. What is different in the case of fluorescent materials is that *only part (not all)* of the absorbed light and ultraviolet radiation is transformed to heat. The remaining absorbed part is transformed and *reemitted as visible radiant energy at longer wavelengths* [Refs. 2.5, p. 15; 5.13]. This transformed energy, reemitted as visible radiation, *adds to* the light normally scattered or transmitted from the material. The result is that, within certain wavelength regions, the light that leaves the surface is often increased sufficiently, in relation to the surrounds, to allow the visual perception of fluorescence. It should be pointed out that fluorescence can occur in some materials without evoking the visual perception of fluorescence [3.9].

Examples of fluorescent materials can be found in which only ultraviolet radiation is transformed to longer-wave visible radiant energy, but the materials of perhaps greater practical interest are those in which both ultraviolet radiation and short-wave visible radiant energy are transformed and produce what is called *daylight fluorescence* [5.14].

An example of what may occur is indicated by the curves in Fig. 5.13 for a transparent fluorescent red film on paper exposed to sunlight [5.15]. Curve I shows the reflectance augmented by fluorescence contributions. At wavelengths shorter than 550 nm most of the light is absorbed. Within the wavelength range 580–680 nm, the curve arches to 165 %, well above the horizontal dashed line that represents the spectral reflectance curve of an ideal white surface. It is clear that *within an appreciable wavelength range* more light is being emitted than is being received. The explanation is that part of the large amount of radiant energy (ultraviolet radiation and light) that is absorbed at wavelengths shorter than 580 nm is transformed and reemitted as light of longer wavelengths, from about 580 nm to over 700 nm.

Curve II (dashed) is interesting because it shows results obtained when sunlight is filtered to eliminate all ultraviolet radiation (wavelengths shorter than 380 nm) before falling on the red film. The area between curves I and II represents the contribution (only about 10 %) made by the sun's ultraviolet radiation.

Curve III indicates the portion that is not absorbed when the light received on the film is sunlight from which both ultraviolet radiation and light at wavelengths shorter than 580 nm have been filtered out to prevent fluorescence. Over the wavelength range 580–700 nm, curve III is an ordinary spectral reflectance curve for the red film treated like a nonfluorescent material. The area between curves II and III is relatively great. It indicates the significant contributions to wavelengths longer than 580 nm caused by transformation of radiant energy absorbed in the visible range between 380 and 580 nm. Knowing this, we should not be surprised to find that the red film appears red when it is illuminated with blue light [5.15].

Spectral reflectance and spectral transmittance curves for *nonfluorescent* materials are not dependent on the spectral power distributions of the light sources used for illumination. They are equally valid for incident light of all different wavelength compositions. The curves for *fluorescent* materials, however, do depend upon the wavelength composition of the illumination. Curve I in Fig. 5.13, for example, is appropriate for the wavelength composition of sunlight in both the ultraviolet and visible domains at which absorption occurs. Figure 5.14 shows the reflectance curves appropriate for two different light sources, for the same colored sample. Curve I for sunlight (from Fig. 5.13) may be compared with curve IV, which is appropriate for illumination provided by an incandenscent-tungsten-filament lamp [Ref. 5.16, p. 36]. The lower peak is explained by the fact that less light and ultraviolet radiation are available at wavelengths shorter than 590 nm in a beam from the lamp (typified by CIE ILL A, Fig. 4.4) than from sunlight (typified by CIE ILL B, Fig. 4.4).

The difference of appearance with fluorescent and nonfluorescent colorants is often so striking that we are tempted to ask what effects might be produced if they were mixed [Ref. 1.39, p. 111]. To understand the problem,

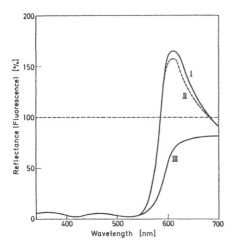

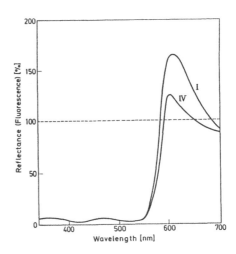

Fig. 5.13. Fluorescent red film on white paper. Spectral reflectance curves for illumination by sunlight (I), by sunlight with the ultraviolet portion filtered out (II), and by sunlight with all radiation below 580 nm filtered out (III). (From [Ref. 5.15, Fig. 7]; reproduced with the permission of The Institute of Physics, Bristol, England)

Fig. 5.14. Fluorescent red film on white paper. Spectral reflectance curves for illumination by sunlight (I) (from Fig. 5.13) and by light from an incandescent lamp (IV) (estimated on the basis of [Ref. 5.16, Fig. 9])

let us consider a mixture of cadmium red (Fig. 5.1) and the fluorescent red colorant (Fig. 5.13) viewed in sunlight. From the above, we note that the fluorescent pigment absorbs radiation in the wavelength range 380–580 nm [radiation that is amply provided by sunlight (CIE ILL B, Fig. 4.4)] and reemits a significant portion of the energy in the wavelength range 580–700 nm in addition to the light scattered in that range. Cadmium red, on the other hand, absorbs about 95 % of incident radiation at wavelengths from 400 to 560 nm, which is dissipated invisibly (as heat). This means that most of the light of wavelengths below 560 nm that reaches cadmium red particles is lost and not available for scattering to neighboring fluorescent pigment particles (Sect. 5.7). Hence the fluorescent emission can be markedly reduced or eliminated, depending on the relative amounts of the pigments. If instead of cadmium red an equivalent amount of madder lake pigment (Fig. 5.1) were used, the fluorescence would also be depressed, but perhaps not quite as much. The absorption of light by madder lake is less (say 65 % absorption) in the wavelength range from 400 to 560 nm, as a result of which, more of the light scattered in this range by the madder lake particles would reach neighboring fluorescent particles.

Optical bleaches (fluorescent brighteners) are used in paper and textile treatment. Such agents absorb ultraviolet radiation (in the wavelength range 300–400 nm) and emit part of it at short wavelengths in the visible range (blue) (mainly 420–430 nm) [Ref. 1.39, p. 111]. The result is that yellowness in a fabric, for example, is "neutralized" (the blue emission adds to the scattered yellow light, its additive complementary color, to produce white light by additive color mixture, Sect. 5.6), and the lightness is enhanced by the increased amount of light that leaves the surface [5.15].

Some fluorescent pigments appear white or whitish in sunlight, but when viewed in darkness while exposed to "black light" (ultraviolet radiation, Sect. 4.1) they glow with highly saturated colors. Minerals that contain fluorescent constituents are often seen displayed in this manner in science museums. Their rich colors often cannot be seen in daylight, because their brightness is much too low in comparison with the brightness of ambient illumination.

5.5 Metamerism and Matching Colors

Matching of colors involves a phenomenon that is fundamental to an understanding of the purpose and method of CIE color specification. It should be considered here, at least briefly. To introduce the idea, let us consider the following example. A small area of green paint has been scraped from a uniformly painted and uniformly illuminated wall. Now it is necessary to repaint the scraped area. House painters and willing artists are able to produce an excellent match even when the pigments in the paint they use are different from the pigments present in the surrounding old paint. But how can the match be a good one? Are not the spectral reflectance curves for the new and old paint films different? Yes, in answer to the second question, the spectral reflectance curves for the two paint films may be very different.

The answer to the first question is related to the fact that the eye cannot identify the wavelength composition (relative spectral power distribution) of light [Ref. 2.5, p. 25]. (In a way, the ear is more analytical than the eye, because it can detect each of the musical tones in a chord.) In fact, *one* color response, for example a particular green, can be evoked by *any one* of a set of stimuli all of which have a different wavelength composition (relative spectral power distribution curve). This set of stimuli is called a *metameric set*. The stimuli in such a set are called *metamers,* and the matching property of such stimuli is called *metamerism*. In the case of matching paint, the *stimuli* (the light that comes to the eye from the two paint films) are matched. (The matched stimuli are metamers.) In reality, paints are not matched by the eye; only stimuli are.

In addition, we should recall that the wavelength composition of the light that enters the eye depends not only on the spectral reflectance curves of the two green paint films but also on the wavelength composition of the light that falls on the paint films (Sect. 5.2, Fig. 5.10). If the wavelength composition of the illumination that falls on the "matching" paint films is changed (for example, from that of light from an incandescent lamp to that of light from a fluorescent lamp), then it is almost certain that stimuli that come from the two films will no longer be metamers and that the perceived colors will be different; the "match" will no longer be good.

The number of stimuli in a metameric set can vary widely. In the case of white light, the number of metamers in the set is very large (Sect. 7.4). This means that light of a very large number of different wavelength compositions can produce the white response. The number in a set is much smaller when the colors of the spectrum are approached. Strictly speaking, for each of the colors of the spectrum there is just one stimulus – monochromatic light – in the metamer set. Thus the wavelength composition is given by one wavelength, for example, 495 nm for a certain bluish green spectral light. Judd estimated that there are more than 10 million metamer sets – that is, more than 10 million colors – that can be discriminated by comparison by an unaided normal eye under suitable viewing conditions [Ref. 2.5, p. 29].

Light sources can be devised whose beams are metameric (metameric illumination [Ref. 2.5, p. 217]). Startling demonstrations can be arranged with them. For example, a beam can be produced that is metameric with daylight, such that a sheet of paper (white in daylight) is also white in the beam, but a lemon (yellow in daylight) is reddish orange in the same beam (Sects. 7.4, 7.13). A demonstration in color is given in [Ref. 2.1, p. 244, Plate XIII] showing the change of color of various objects under two metameric white illuminations.

5.6 Additive Color Mixture

Just as the term "matching paint" is inaccurate, because in reality a stimulus light, not paint, is matched, so the term "color mixture" is inaccurate, because it is a stimulus light, not a response color, that is mixed. There would be some merit in replacing "color mixture" by a more scientifically correct term like "color-stimulus synthesis" [Ref. 1.20, p. 115], just as there would be in replacing the usual commercial term "ice cream" by a term closer to the mark like "frozen milk product". But I am sure that it is more important both to be aware of the facts and to maintain general communication by using universally adopted terms. With this in mind, let us proceed to the subjects of additive color mixture, subtractive color mixture, and color mixture by averaging [5.17].

Additive color mixture occurs when light of different colors from two or more sources is combined (added together) before it reaches the eye. A helpful way to demonstrate the effect is to project two beams of colored light onto a white wall so that the two disks of light are superimposed and the combined light from the two beams is scattered from the wall. Because the two beams are added together, the energy in the combination is equal to the sum of the energies of the two initial beams. Usually the effect is apparent in a resulting enhanced brightness when one disk is superimposed on the other (but there are exceptional situations [Ref. 5.18, p. 144]). If the perceived hues of the two initial beams are different, the resulting combined beam will generally be perceived to have an intermediate hue. Thus, if the beams are red and green, the hue of the superimposed disks may be yellow green, yellow, or orange, depending on the relative intensities of the initial beams. If, however, the hues are sufficiently different [at opposing positions on the color circle (Sect. 5.10)], for example a red and a blue green, then by appropriate adjustment of the relative light intensities it is possible to produce a hueless (white) response. In this case, the two original colors are called *complementary colors* [or, more precisely *additive complementary colors* (Sects. 6.2, 7.2)]. In general, the saturation of the color of the combined beam is less than that of at least one of the original beams. With complementary colors, of course, the saturation of the color of the combined beam can be as low as zero (white light).

5.7 Subtractive Color Mixture

As described, additive color mixture can be demonstrated by the combination of two or more beams of light of different hues to produce a beam of yet another hue. *Subtractive color mixture,* on the other hand, can be performed with one beam, from which energy in different amounts at various wavelengths is removed (by absorption) by two or more different colorants such that the resultant beam produces a different hue.

A beam of sunlight, for example, can be passed in succession through two pieces of colored glass (light filters), one yellow and the other green, to produce light of hue yellow green. When sunlight falls on the yellow glass, the yellow light transmitted has a wavelength composition that shows that most of the shorter-wavelength light (blue to green) is absorbed in the glass and that much of the intermediate-wavelength light (yellow green) and most of the longer-wavelength light (yellow, orange, and red) are transmitted. On the other hand, the green light produced when sunlight is passed through the green glass has a wavelength composition that shows that relatively large amounts of blue green, green, and yellow green light are transmitted and that most of the blue, yellow, orange, and red light is absorbed. From this, it is clear that when the beam of light is passed through one glass and then through the other, only the yellow green light emerges, because practically all light at other wavelengths is absorbed (subtracted out). Thus subtractive color mixture occurs when filters are "mixed" (placed in tandem) and light passes through them successively.

The same process occurs, although probably not exclusively, when paints are mixed, for example yellow and green oil paints. A ray of light that enters the paint film becomes scattered and passes in diverse directions through a mixture of yellow and green pigment particles, eventually emerging as yellow green light. Subtractive color mixture can also be demonstrated by passing a beam of light through a solution of two dyes. Again, each colorant (yellow or green dye) selectively absorbs light in its own way and the emerging beam is composed of wavelengths (yellow green) that largely escape both absorption processes. Because energy is removed from the beam by absorption, the intensities of the emerging wavelengths in subtractive mixture are always diminished. The effects can also be explained by using spectral transmittance curves for the green and yellow filters and by using spectral reflectance curves for the green and yellow pigments (Sects. 5.2,3) [Note 5.1].

5.8 Color Mixture by Averaging

Additive color mixture is the process of combining light beams of different colors before they reach the eye. However, beams can be combined *in* the visual process, as when light beams of different colors stimulate the same portion of the retina but without superposition. This can occur when the details in the retinal image produced by an array of tiny beams of different colors are more minute than is the "weave" or "mosaic" of receptor cells and interconnected nerve cells that respond to the array. In such a case, the different colors are not resolved and a kind of retinal mixture or blending takes place, which is sometimes called *spatial averaging* [Ref. 1.20, p. 115].

A combination can also occur in the visual process when a rapid succession of flashes of light of alternating colors falls on an area of the retina. If the change is too rapid for the visual process to keep pace, retinal temporal mixture (*temporal averaging*) results [Ref. 1.18, p. 66; 1.20, p. 115]. In both cases, there is a "mixed" response: the mixed color is seen.

These two kinds of combination (spatial and temporal mixture) are called *color mixture by averaging*. Thus, a printed paragraph of words in black letters on white paper becomes a gray area when viewed from a sufficient distance; the gray is the result of color mixture (black and white) by averaging (spatial mixture). The same occurs when juxtaposed dots of three different colors are viewed on a television screen, or in halftone printing on paper (Sect. 7.9). The same may occur when a pointillistic painting is viewed from so great a distance that the dabs of paint of different color are not individually distinguished. As the painting is approached and the individual color dabs begin to become distinguishable, other effects occur (Sect. 11.11) [Refs. 2.5, p. 214; 2.6]. A good example of rapidly alternating stimuli leading to color mixture by averaging (temporal mixture) is a rapidly turning disk (sometimes called a *Maxwell disk*) whose surface has sectors of different colors [Refs. 1.17, p. 17; 1.18, p. 66; 2.3, p. 90]. Thus a disk whose surface is covered by black and white sectors will appear gray while turning.

The difference between additive color mixture and color mixture by averaging is implied by the two terms. In additive color mixture, the energy of the two combined beams is the result of adding the energy of the two initial beams, and the brightness is usually (but not always) increased. In color mixture by averaging, the effective energies are area averaged or time averaged in the visual process, and a kind of average brightness is generally produced. The terms *optical mixture* and *visual mixture* are commonly used in reference to color mixture by averaging.

5.9 The Primaries

To many people, the terms *primaries* and *primary colors* suggest bright red, yellow, and blue colors. With paints of these colors, and of white and black in addition, mixtures can be made that produce colors of a wide range of hues, lightness, and saturation. Because it is desirable here to consider color mixture and matching in a broader sense, the concept of the primaries as stimuli (light) will be further elucidated.

A basic requirement of a set of three primaries is that no combination of any two of them matches the third [Refs. 1.20, p. 119; 1.39, p. 73]. To have particular utility, the qualification is usually added that the three primaries be selected so that the gamut of colors obtainable by mixing them includes all hues and is as large as is practical.

A very large gamut of colors (Hardy-Wurzburg gamut, Sect. 7.3) that includes all hues may be produced by mixtures (additive color mixture) of varying proportions of monochromatic light of three wavelengths: 700 nm (red), 535 nm (yellowish green), and 400 nm (bluish purple, or violet) [Ref. 2.1, p. 238]. The color gamut includes all the purples and most of the reds, oranges, and yellows; only the high-purity greens and blues are excluded (Sect. 7.3). Three such stimuli are called *additive primaries*. In a less specific way, the additive primaries are ordinarily said to be stimuli that produce the responses *red, green*, and *blue*.

In some situations the *subtractive primaries* are of importance. The colors produced by each of the three subtractive primaries are complementary to red, green, and blue, – namely, *cyan* (blue green or turquoise), *magenta* (purplish red), and *yellow*, respectively. Cyan, magenta, and yellow pigments and dyes are used in color photography and in the four-color (including black, to improve blackness and definition) printing process. In these cases, there is an advantage in having to deal with only three or four dyes or inks. Admittedly, painters and designers, who use tubes or jars of paint of a great variety of colors, may not find the subject of subtractive primaries directly relevant to their work, but, as will be shown in the following chapters, the additive primaries are basically relevant in art and design.

5.10 Color Circles

Color circles of the type familiar to artists generally present color samples in the sequence of spectral hues, as found in a rainbow. A color circle is completed by inserting the nonspectral hues (purples and purplish reds) between violet and red. Colors can be selected to form a color circle such that pairs of complementary colors [additive complementary pairs (Sect. 7.2) or afterimage complementary pairs (Sect. 11.7)] are directly opposite each other [Ref. 1.20, p. 116].

A circle of six members may be formed that has samples that represent the three additive primary colors and their additive complementary colors, the subtractive primary colors (Fig. 5.15 and Plate I). Intermediate complementary-color pairs may be introduced to increase the number of colors to 12, 24, 48, 96, or 192. But in a circle that contains 192 colors, the difference between the hues of adjacent colors is hardly perceptible (Sect. 3.2).

F.J. Gerritsen, opposed to the old practice of teaching art students that the primary colors are blue, red, and yellow, has proposed a color circle of additive complementaries as a teaching aid [5.19,20]. Color circles containing 12 and 15 additive complementary pairs have served as the basis of several editions of an important color-sample atlas, the *Color Harmony Manual* [1.32] (Sect. 8.5).

A six-member color circle consisting of three afterimage complementary-hue pairs is properly called a *Goethe color circle*, for Goethe is credited with

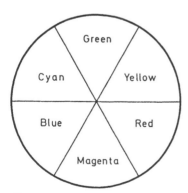

Fig. 5.15. A six-member color circle. Opposing additive complementary pairs (Sect. 7.2)

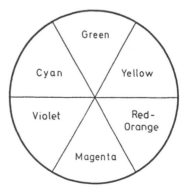

Fig. 5.16. A Goethe color circle. Opposing afterimage complementary pairs (Sect. 11.7)

its introduction [Refs. 1.2, pp. 41, 49; 5.3, p. 664]. Examples of such a circle are shown in Fig. 5.16 and Plate I. Goethe determined his "physiological complementaries" [Refs. 1.1, Sect. 3, p. 2, Sect. 47, p. 20; 5.21], the term he used to refer to afterimage complementary pairs, in his studies of color contrast (1793). They were red/green, blue/orange, yellow/violet [Refs. 1.1, Plate I; 1.2, pp. 41, 49; 5.22, p. 205; 5.23] .

If we wish to make a color circle of a relatively large number of colors (24, 48, or 96, for example), we should consider the possibility of having an approximately visually uniform hue sequence with afterimage complementary pairs in diametrically opposing positions (Sect. 11.7). (An arrangement of equally spaced hues with additive complementary pairs in opposing positions has not been found [5.21].) Furthermore, it should be noted that a circle based on afterimage complementary pairs has a more balanced appearance than one based on additive complementary pairs [Ref. 5.3, p. 664]. In the latter, there appear to be an excess of blue green hues and a deficiency in blue and red hues.

Another type of color circle involves the four chromatic psychological primaries (unitary red, yellow, green, and blue) positioned at equal (90°) intervals (Sects. 3.2, 8.7, 11.2). The binary intermediate hues yellow red (orange), red blue, blue green, and green yellow are located midway between the corresponding unitary hues. This type of circle is embodied in the Swedish Natural Colour System (NCS) described in Sect. 8.7 and in the Hurvich-Jameson HBS System (Sect. 11.3). NCS color chips and papers are available that may be used for the construction of circles containing 4, 8, 16, 32, ... members.

Series of color samples whose hues differ by equal, or approximately equal, perceptual amounts are found in three well-established color-sample collections: *The Munsell Book of Color* (Sect. 8.4), The *DIN-6164 Color Chart* (Sect. 8.6), and the *OSA Uniform Color Scales* (Chap. 9). In making a color circle using color chips or papers from these collections, one can make selections taking full account of two other controlled variables. For example, in the Munsell collection one can select chips for a hue circle having a specific Value and Chroma (Sect. 8.4). A complete circle of approximately equally spaced Munsell hues would contain 40 chips.

6. Color Specification (CIE)

6.1 Light and Color: Other Definitions

The subject of color measurement, *colorimetry,* is in the domain called psychophysics, which lies between the domains of psychology, physics, physiology, and chemistry [Ref. 4.18, p. 40]. In the early 1930s, colorimetry was put on a universally accepted precise quantitative basis. The scheme, however, required a redefinition of basic terms.

As stated in Sect. 4.1, we commonly define light as visible radiant energy [Note 4.1]. In psychophysics, however, a definite distinction is made between light and visible radiant energy. Here the meaning of the term "visible radiant energy" is retained; it is radiant energy in the range from 380 to 780 nm and the stimulus of vision [Ref. 4.18, p. 13].

Light, on the other hand, has been defined in psychophysics to take account of a human observer's awareness: light is "the aspect of radiant energy of which the human observer is aware through the agency of his eyes and the associated nervous system" [Ref. 4.18, p. 40]. The distinction is clear if we consider that we are not equally aware of visible radiation received in equal amounts at 381 nm (barely visible) and 555 nm (of maximum visibility). Thus, in psychophysics "visible radiant energy" refers to all radiation in the visible range, and "light" refers to the same radiation but with the relative magnitude of its effectiveness in producing vision taken into account.

With light considered in the psychophysical sense, we can proceed to the psychophysical definition of "color", the "psychophysical color" mentioned in Sect. 3.6. The word "color" in psychophysics denotes a characteristic of the *stimulus* – that is, of the visible radiant energy. (This is closer to the layman's concept that light is colored). It takes into account both the radiant energy that reaches the eye and a standard observer who has typical normal color vision and, hence, makes typical use of the radiation that produces vision. The Committee on Colorimetry of the Optical Society of America, having adopted the psychophysical concept of color, reported, "This course seems to be amply justified on purely philosophical grounds, but, if less academic justification is desired, the purely practical considerations are fully sufficient" [Ref. 4.18, p. 13].

Often, colors can be measured by finding a match to one of a series of standard samples (such as printed papers, dyed fabrics, and paint swatches or chips) under standardized conditions of viewing. For greater accuracy, devices called *colorimeters* can be used. In one type of colorimeter, the field of view contains the color sample and the comparison color. The latter is varied by three kinds of adjustments until a match is found. The color is then expressed in terms of three numbers that, in the case of some instruments, represent directly the internationally accepted *CIE tristimulus values* (Sect. 6.3) or else can be converted to them. Photoelectric colorimeters operate automatically; in them the human eye is replaced by a photoelectric cell whose spectral responses are adjusted to mimic human vision.

A colorimeter provides a direct measurement of color. There is, however, an indirect method that provides the CIE tristimulus values; it is more precise and quite extensively employed. The method involves the use of a spectrophotometer to obtain a spectral reflectance curve for an opaque sample (Sect. 5.2) or a spectral transmittance curve for a transparent sample (Sect. 5.3). With use of the curve, the CIE tristimulus values can be calculated in a routine manner for the sample in a selected kind of illumination (for example, illumination typified by CIE ILL C or CIE ILL D_{65}). The possibility of human error in making direct color measurements is avoided in this procedure. Imperfections of photoelectric-cell adjustment in colorimetry are also bypassed. This indirect approach is of particular interest because the scheme established for it provides access to the structure that underlies the CIE tristimulus values, which are basic to the precise specification of colors. Some of the fundamental ideas involved are taken up in Sects. 6.2, 3, and 5.

6.2 The Chromaticity Diagram: An Introduction

Let us begin with the Maxwell triangle, the framework of the chromaticity diagram, which is in universal use in commerce, industry, and science and is now appearing in literature intended for artists and designers. The subject is easily approached by considering the example of three beams of light of short, medium, and long wavelengths used as a set of three additive primaries (blue, green, and red). Different colors are produced by superimposing the disks of the three beams projected on a white wall (additive color mixture) and varying the amount of light in each beam. If the color produced by a fourth beam of light is within the gamut of colors that can be produced by mixtures of the three beams, then the color can be specified by the amounts of each of the three beams (primaries) required to match it.

A set of three additive primary colors and the complete gamut of colors obtainable by mixing two or three of them can be represented on a kind of *mixture diagram,* an equal-sided (equilateral) triangle, the so-called *Maxwell triangle,* named after the Scottish physicist James Clerk Maxwell (1831–1879), who employed it in his basic work on color. The three primaries are

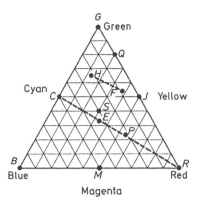

Fig. 6.1. Chromaticity diagram or Maxwell triangle (equal-sided triangle)

assigned to points at the corners of the triangle. The gamut of colors of all possible mixtures of the particular primaries is represented by points on the three sides of the triangle and by points within it (Fig. 6.1).

The representation of psychophysical color by the triangle is partial. The part that is represented is called the *chromaticity,* and, indeed, it is now much more common to refer to the triangle as a *chromaticity diagram.* Chromaticity is the *quality* aspect of psychophysical color; it is a composite representation of approximate equivalents of psychological hue and saturation. The part not included on the diagram, the *quantity* aspect of psychophysical color, is the effective amount of light – that is, the amount of psychophysical light defined in the previous section. It is the amount sensed in the visual process. Because the eye's efficiency in responding to a given amount of radiation varies from zero at the limits of visibility (380 and 780 nm) to a maximum at 555 nm (Sect. 4.7), the psychophysical amount [called the *luminance* (Sect. 6.3)] is taken as the physical amount weighted by the eye's efficiency.

A condition imposed on the selection of the colors to serve as primaries is that, when the three beams are combined in psychophysically equal amounts, a white disk is produced on a white screen or wall. The chromaticity of such an *equal-energy white* is represented by point E at the center of the chromaticity diagram (see also Sect. 6.3).

Although the equal-sided triangular chromaticity diagram is employed rarely, it is instructive to consider it a bit further before we examine the diagram that is in present-day use. The equal-sided triangular diagram may be presented on triangular-coordinate paper, which is subdivided by three superimposed sets of parallel lines (Fig. 6.1). The three sets are shown separately in relation to the triangle in Figs. 6.2–4.

It is helpful to borrow the CIE symbols X, Y, and Z; these represent the CIE tristimulus values, which are defined for the particular set of primaries discussed in Sect. 6.3. Here, however, they are taken to represent the amounts of the three primaries now being discussed (X, amount of red; Y,

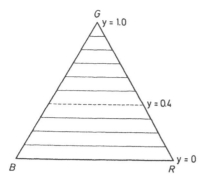

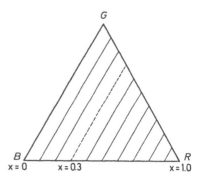

Fig. 6.2. Lines of constant y (fractional amount of primary green)

Fig. 6.3. Lines of constant x (fractional amount of primary red)

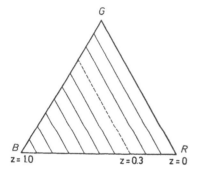

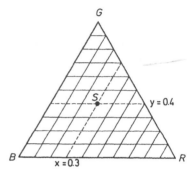

Fig. 6.4. Lines of constant z (fractional amount of primary blue)

Fig. 6.5. Location of point S

amount of green; and Z, amount of blue) needed in a mixture to match a color within the gamut. The relative proportions or *fractional amounts* (given by x, red; y, green; and z, blue) clearly represent the quality aspect of psychophysical color and, as we shall see, they locate the chromaticity by a point on the chromaticity diagram.

If the tristimulus values are $X = 60$, $Y = 80$, and $Z = 60$, then the fractional amount of primary red x, for example, is the amount of red (60) divided by the total amount (200), which is 0.3. The calculations similarly yield 0.4 for y and 0.3 for z [Note 6.1]. The chromaticity of the color can now be represented on the diagram.

Figure 6.2 shows a series of parallel lines along each of which y does not vary, for values of y from zero at the base of the triangle (0 % primary green in the mixture) to 1.0 at the vertex of the triangle (100 % primary green). Because $y = 0.4$, the sought point must be located somewhere on the line labelled $y = 0.4$, shown as a dashed line. Figure 6.3 presents similar information for the primary red. Because $x = 0.3$, the point must be located on the line labelled $x = 0.3$. The point can be located on both lines simultaneously only at their intersection (Fig. 6.5). By use of Fig. 6.5, the sought point S can be transferred to the chromaticity diagram (Fig. 6.1). It is clear that the information $z = 0.3$ and the point's location on a line labelled $z = 0.3$ in Fig. 6.4 are not needed. *Only two* fractional amounts are necessary to locate the point and to specify the chromaticity; the two commonly employed are x and y. When values are given for x and y, the value for z can always be obtained by subtracting the sum of the values of x and y from 1.0.

The chromaticity of the color just discussed is written as $(x = 0.3, y = 0.4)$ or, more commonly, as $(0.3, 0.4)$, – that is, (x, y). In the case of white (equal-energy white), for which the amounts of the primaries required are equal, the fractional amounts x, y, and z are obviously each equal to 1/3, or 0.333, and the chromaticity is specified by $(0.333, 0.333)$. The central location of point E (Fig. 6.1) can be checked by plotting the point by use of the procedure just described.

Point Q that is plotted on one side of the diagram shows the chromaticity of a mixture of two primaries, red and green. For a mixture of equal amounts of primaries red and green (X and Y are equal, and Z is zero – there is no primary blue), fractional amounts x and y are equal, – that is, 1/2, or 0.5 – and the chromaticity is $(0.5, 0.5)$. This chromaticity is represented by point J on the diagram and corresponds to yellow. Similarly, equal amounts of primary red and primary blue result (when the amount of primary green Y is zero) in a light beam that produces the color magenta $(0.5, 0.0)$ (point M); and equal amounts of primary blue and primary green result (when the amount of primary red X is zero) in the color cyan $(0.0, 0.5)$ (point C).

The sequence of the colors around the triangle (Fig. 6.6) is the same as that in the six-member color circle showing additive complementary pairs (Fig. 5.15). *Additive complementary color pairs* are found by drawing straight lines across the diagram, through point E, such as the dashed line that connects red and cyan (Fig. 6.1). Constructing a line from point

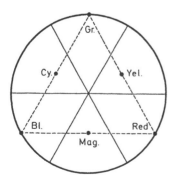

Fig. 6.6. Superposition of the Maxwell triangle and a six-member color circle of additive complementary pairs. (From Fig. 5.15)

51

Q (yellow green) through point E to the opposite side would show that the complementary color is a purple. More will be said about additive complementary colors in Sect. 7.2.

The colors produced by mixtures of light are of maximum saturation when the chromaticities of the mixtures are located on the sides of the diagram. The chromaticity of white, for which the perceived saturation is the minimum (zero), is located at the center E of the diagram. The variation of saturation can be demonstrated by combining complementary red and cyan beams of different relative amounts, as shown by the dashed line connecting R and C in Fig. 6.1. Let us start at point R which corresponds to the chromaticity of the red color of the beam at full intensity with the cyan beam shut off. Then let us diminish the intensity of the red beam progressively and increase the intensity of the cyan beam. The perceived saturation of the red beam is found to decrease while the chromaticity passes continuously through P (pink) to point E. Beyond E the hue changes to cyan and the saturation of the colors increases to a maximum at C. The dashed line that connects R and C is a *mixture line;* it represents the gamut of chromaticities that can be produced by mixing the two beams.[1] Another mixture line is shown between points H and F, representing a gamut of mixtures of a yellow and a green; white is not included in the gamut (the line does not pass through point E).

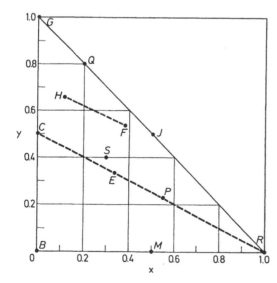

Fig. 6.7. Chromaticity diagram or Maxwell triangle (right triangle)

---

[1] There is some variation of the perceived red and blue hues along the mixture line passing through point E, but this variation may be disregarded to simplify the discussion here and in Sect. 6.3. Some idea of the extent of variation may be gained from the plots of lines of constant Munsell Hue radiating from the point for CIE ILL C on the CIE (x, y) chromaticity diagram (Figs. 12.1–9).

The equal-sided chromaticity diagram, with its triangular grid showing x, y, and z is useful as an introduction to the subject, but it is rather awkward to employ in practice. It is much more convenient to use a right triangle that has two equal legs and an ordinary square grid showing x and y (Fig. 6.7). Such a triangle can be used because, as noted above, it is not necessary to include z. All points that appear in Fig. 6.1 are also shown in Fig. 6.7. The two mixture lines have been transferred as well. As will be indicated later (Sect. 7.3), it is an important fact that the mixture lines (additive color mixture) are straight.

6.3 The CIE Chromaticity Diagram

It was stated in the previous section that if the color produced by a beam of light is within the gamut of colors that can be produced by mixtures of three beams of primary colors, then the color can be specified by the amounts of each of the beams required to match it. But what can be done if the color is not within the gamut? It is well established that no three primaries can, by their mixture, produce all colors. This problem can be solved by adding one of the primary beams to the beam whose color is being measured, to bring it within the gamut. In such a case, the amount of the primary added is reported as a negative number.

From 1928 to 1930, separate laboratory investigations by W.D. Wright and J. Guild [6.1] obtained data which, when transformed to a common basis, presented the amounts (positive and negative) of three monochromatic primaries (435.8, 546.1, and 700 nm) needed to match the colors of the spectrum [Refs. 1.20, p. 129; 3.14, p. 264; 4.4, p. 10; 6.2, p. 99]. Their data, along with the previously obtained brightness sensitivity data (Sect. 4.7), were adopted in 1931 by the CIE to characterize the visual response of a typical normal viewer called the *CIE 1931 standard observer*. The data served to provide the numerical basis of the internationally accepted CIE method for color specification.

Although it is possible to deal with negative numbers in colorimetry, for various practical reasons it was decided by the CIE to produce a scheme in which negative amounts of primaries do not arise [Refs. 4.4, p. 11; 6.2, p. 101]. Because this is not possible with real primaries, it was necessary to invent primaries: the CIE imaginary primaries. These imaginary primaries are relevant to typical color vision because they are related to the laboratory data by means of mathematical transformations. The gamut of colors produced by mixture of the imaginary primaries includes *all real colors*. The gamut also includes imaginary colors; these are segregated and ignored.

The amounts of the three imaginary primaries necessary to match unit energy of each wavelength in the visible spectrum are recorded as columns of data in a table; the three sets of data, called *color-matching functions,*

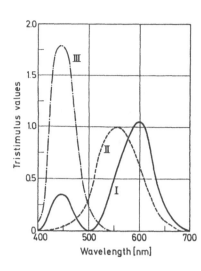

Fig. 6.8. CIE 1931 color-matching functions of CIE imaginary additive primaries [red (*I*), green (*II*), and blue (*III*)]

are plotted in Fig. 6.8. The color-matching functions define the CIE 1931 standard observer.

Thus, from the curves can be read the relative amounts of the imaginary primaries required in additive color mixture to match the colors of monochromatic light (spectrum colors) at any wavelength within the range 400–700 nm. The curves were devised so that when equal amounts of the three primaries are combined their mixture matches equal-energy light *E* (Figs. 4.3 and 4). The color-matching functions are used in calculations to provide the *CIE tristimulus values X, Y, Z*, which represent the relative amounts of the imaginary primaries required to match any color by additive color mixture [6.3].

An important characteristic of the imaginary primaries red and blue is that they have zero luminance [Ref. 6.2, p. 104]. This is a simplification provided by the mathematics that underlies the CIE system. All of the luminance is assigned to the imaginary green primary. The imaginary red and blue primaries were designed so that the color-matching function for imaginary green would be identical to the relative luminosity curve for normal eye brightness sensitivity (Sect. 4.7). To understand the significance of this characteristic, note that, although the imaginary primaries are measured in the same units (from the fact that the amounts required to produce white by a mixture of the three beams are equal), we do not know what the units are. As a result, the calculated tristimulus values (for example, $X = 1300$, $Y = 1000$, and $Z = 1100$) are only relative; they are not absolute values. But because Y is given an alternative meaning, luminance, a separate measurement may be made that provides an *absolute* number, for example, $Y = 200$ (luminance units). Then, to specify the color the tristimulus values may be adjusted in proportion to give $Y = 200$, thus: $X = 260$, $Y = 200$, and $Z = 220$. However, it is conventional, in the case of lights, to report the tristimulus values on the basis of $Y = 100$ (hence, $X = 130$, $Y = 100$, and $Z = 110$) and to quote the luminance (200) separately.

The technical definition of luminance is avoided here; it is sufficient to regard it as an amount of light, where light is considered in the psychophysical sense (Sect. 6.1). Thus an "amount of light" is an amount of radiation weighted by the eye's efficiency in responding to the radiation (Sects. 4.7, 6.2). Luminance is very commonly taken as a correlate of the perception of brightness, but, as such, it is only approximately valid [4.3, 6.4].

The foregoing discussion is concerned with color measurement in the case of direct light from luminous sources such as lamps. But what is done in the measurement of the colors of opaque and transparent nonluminous objects? The situation is much the same. The color measurement is performed on the light being diffusely reflected from (scattered by) an illuminated opaque surface or being transmitted from a transparent surface. The color depends on the reflectance (or transmittance) characteristics of the object *and* on the wavelength composition of the light that illuminates it. Thus, a surface color may be specified by the tristimulus values X, Y, Z, and a standard illuminant such as CIE ILL D_{65}. But, for opaque and transparent materials, Y has a *relative* meaning. For an opaque material, Y is the *luminance factor* (or *luminous reflectance*), which is the luminance of the surface relative to the luminance of an ideal white surface that has the same illumination and angle of view [6.5]. Or stated another way, the luminance factor is the "amount of light" reflected from the surface divided by the "amount of light" received by the surface. For a transparent material, Y refers to *luminous transmittance*, which is the "amount of light" transmitted through the material and leaving one surface divided by the "amount of light" entering at its opposite surface.

Although colors can be specified by the CIE tristimulus values X, Y, and Z, it is rarely done. (Methods for calculating the tristimulus values are given in [3.14, 4.4].) It is more meaningful to employ either *chromaticity* (x,y) or dominant wavelength and purity (discussed in the next section) than to use X and Z. A CIE color specification based on chromaticity is written CIE(x, y, Y), and an identification of the illuminant is added if the object is nonluminous. The values of x and y, which are the fractional amounts of the imaginary red and green primaries in the mixture, are very easily calculated from the CIE tristimulus values, as described in Sect. 6.2 [Note 6.1].

As in the case of real primaries (Sect. 6.2), the chromaticities of the imaginary primaries occupy the corners of the triangular diagram, and the chromaticity of any color (real or imaginary) that results from their mixture is represented by one point plotted within the triangle or on one of its three sides. White produced by an equal mixture of the primaries is represented by its chromaticity at the center E, as it was in Fig. 6.7.

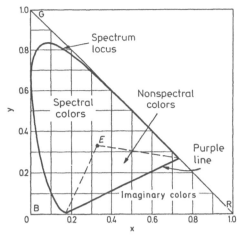
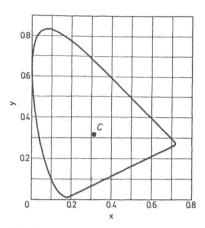

Fig. 6.9   Fig. 6.10

Fig. 6.9. CIE 1931 (x, y) chromaticity diagram or Maxwell triangle (right triangle) based on the three imaginary primaries (R, G, B)

Fig. 6.10. CIE 1931 (x, y) chromaticity diagram as ordinarily presented. A chromaticity point for an illuminant is frequently presented, usually C (CIE ILL C), D_{65} (CIE ILL D_{65}), or A (CIE ILL A). Illuminant E is less frequently found

The chromaticities of all real colors fall within a tongue-shaped area or on its borders (Fig. 6.9). The area outside the tongue-shaped area is the site of points that represent chromaticities of imaginary colors; it is therefore of no practical interest. For this reason, the outside area and the sides of the triangle are disregarded; in practice, only the tongue-shaped area is presented (Fig. 6.10). This is the internationally accepted *CIE 1931* (x,y) *chromaticity diagram.*

Now let us consider more closely the general structure of the CIE diagram while keeping the original triangle in mind (Fig. 6.9). The top of the tongue-shaped area is the site of greens; the bottom-left area, blues; and the bottom-right area, reds.[2] The chromaticities of all colors produced by monochromatic light are located along the curved line outlining the tongue-shaped area. This curved line is called the *spectrum locus.* Because monochromatic radiation is by definition light of a single wavelength, the wavelength scale is sometimes indicated along the spectrum locus (Fig. 6.12). The straight line (called the *purple line*) that borders the bottom of the tongue-shaped area connects the chromaticities for red (wavelength 700 nm) and blue (380 nm) and represents the chromaticities of their mixtures, which produce certain reds and the full range of purples, all of maximum saturation. As mentioned earlier (Sect. 4.3), colors that have purple hues, or hues

---

[2] Some idea of the distribution of hues on the CIE 1931 (x, y) chromaticity diagram may be gained from the locations of color samples shown in Plate II.

of red associated with the purple line, are called *nonspectral colors;* all other chromatic colors are called *spectral colors* (Fig. 6.9).

The location of a chromaticity point reveals some information about the saturation of the perceived color – namely, the closer the point is to the spectrum locus or to the purple line, the higher the saturation. Zero saturation is found when the point is in the central region (around E). This has been discussed in Sect. 6.2 for the chromaticity diagrams of Figs. 6.1 and 7, and more will be said about this topic in Sect. 6.4.

The 1931 chromaticity diagram is based on test data for a narrow *angle of vision* (2°) and is considered suitable in colorimetry for angles from 1° to 4° [Ref. 5.16, p. 25]. Actually, its use is recommended without restriction of the angle of vision [6.6]. An angle of 4° projected from the eye includes a disk 17 cm in diameter located at a distance of 2.5 m; an angle of 1° includes a disk of 4.4 cm diameter at the same distance. For an angle less than 1°, images obtained by looking directly at an object fall on the retina of the eye within a region called the *fovea,* which is the region that permits the sharpest vision (Fig. 2.1) [Ref. 2.3, p. 117].

Because colorimetry often involves large angles of vision, in 1964 the CIE recommended a supplementary set of data, derived from observations made using a 10° angle of vision, for optional use for visual fields subtend-

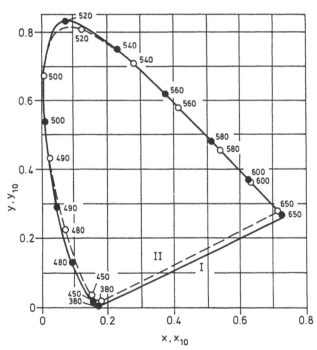

Fig. 6.11. CIE 1931 (x, y) chromaticity diagram (I) and CIE 1964 (x_{10}, y_{10}) chromaticity diagram (II). (Based on [Ref. 1.18, Fig. 2.17]; reproduced with the permission of John Wiley & Sons Inc., New York)

ing 4° or more [Ref. 4.4, p. 10]. The supplementary set of data defines the *CIE 1964 supplementary standard observer.* An angle of 10° includes a disk of diameter 44 cm at a distance of 2.5 m. There is a difference between the 1931 and 1964 chromaticity diagrams (Fig. 6.11) because, for a 10° angle, the image on the retina extends farther beyond the edge of the fovea, and consequently somewhat different color responses are produced. The chromaticity for the equal-energy source E is the same (0.3333, 0.3333) on both chromaticity diagrams, but the chromaticities for the CIE illuminants are different (Table 7.6).

The notation in the 1931 system is X, Y, Z and x, y, z; that for the 1964 system is X_{10}, Y_{10}, Z_{10} and x_{10}, y_{10}, z_{10}. For color specifications based on the 1931 system, it is recommended to write CIE $1931(x, y, Y)$, and for those based on the 1964 system to write CIE $1964(x_{10}, y_{10}, Y_{10})$. Most of the applications discussed in this book make use of the 1931 system. An example of a specification of the cadmium red pigment discussed in Sect. 5.2 is CIE 1931 (0.5375, 0.3402, 0.2078) CIE ILL C. The luminance factor Y is given as a fraction; it may also be given as a percentage (e.g., 20.78 %). The illuminant is cited, as it must be for the specification of the color of an illuminated object. When there is no doubt that the date 1931 is intended, it is usually omitted in a specification.

6.4 Dominant Wavelength and Purity

There is an acceptable way to specify color that is more descriptive than CIE (x, y, Y); it is favored in certain industries. The notation is given by CIE (λ_D, p_e, Y), where λ_D is the Greek letter lambda (with a subscript) used as the symbol for *dominant wavelength,* p_e is the *excitation purity,* or simply *purity,* and Y, as before, is either the luminance, the luminance factor, or the luminous transmittance.

An idea of what dominant wavelength and purity are can be gained from the examples of the cadmium red and madder lake pigments discussed earlier (Sect. 5.2). The method requires the choice of a reference point on the chromaticity diagram, which characterizes the illumination that is applicable, in this case CIE ILL C (daylight). Point C in Fig. 6.12 represents the chromaticity of CIE ILL C (Table 7.6), and point P represents the chromaticity of the cadmium red pigment. A straight line drawn from C through P intersects the spectrum locus at a point that represents the chromaticity of a color produced by light of a single wavelength (monochromatic radiation). Since the wavelength of monochromatic radiation specifies its own hue, under controlled conditions (Table 4.2), it can serve as an indicator of the hue of the pigment's color, because the latter can be matched by a mixture of the monochromatic radiation and the white light from the illu-

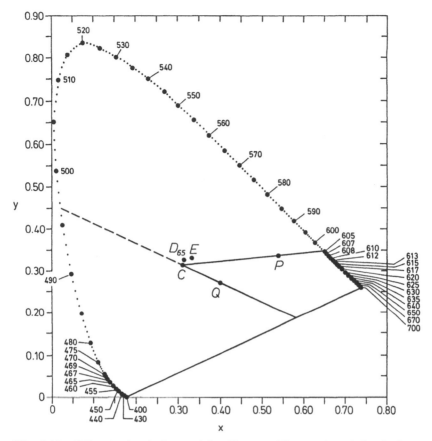

Fig. 6.12. CIE 1931 (x, y) chromaticity diagram. The wavelength [nm] of monochromatic light is indicated along the spectrum locus for use in determining the dominant, or complementary, wavelength of a color (see also Fig. 7.2). The chromaticity point for a standard illuminant (CIE ILL C, CIE ILL D_{65}) or that for equal-energy light (E) may be used as the reference point

minating source. Used in this way, the wavelength is called the *dominant wavelength* (in this case 605 nm).

To determine the purity p_e, measure the distance between points C and P and divide it by the total length of the line measured from C to the spectrum locus. The quotient, or fractional distance of P on the path from C to the spectrum locus, is the *purity*. Clearly, if the position of P on the line were varied from C to the spectrum locus, purity would increase from zero to 100 %. For the cadmium red pigment the purity is 0.673 or 67.3 % and the specification is given by CIE 1931 (λ_D = 605 nm, p_e = 67.3 %, Y = 0.208), CIE ILL C.

A second example is needed to illustrate what is done in the case of nonspectral colors. Point Q represents the chromaticity of the color of the madder lake pigment (Fig. 6.12). In this case, the line drawn from reference point C through Q intersects the purple line. The purity p_e is determined as described above: the distance from C to Q is divided by the length of the line drawn from C to the purple line. Here the purity is found to be 0.329 or 32.9 %.

But how does one represent the hue? Monochromatic radiation cannot produce purples and purplish reds. For nonspectral colors, the line is extended in the opposite direction (see the dashed portion in Fig. 6.12) to the spectrum locus, and the *complementary wavelength* λ_c is reported, in this case 496.5c nm (a "c" is always affixed to avoid confusion). The specification for madder lake is CIE 1931 (λ_c = 496.5c nm, p_e = 32.9 %, Y = 0.336), CIE ILL C.

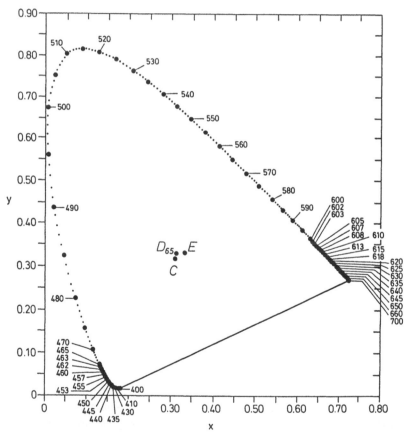

Fig. 6.13. CIE 1964 (x_{10}, y_{10}) chromaticity diagram. The wavelength [nm] of monochromatic light is indicated along the spectrum locus for use in determining the dominant, or complementary, wavelength of a color. The chromaticity point for a standard illuminant (CIE ILL C, CIE ILL D_{65}) or that for equal-energy light (E) may be used as the reference point

60

Figures 6.12 and 13 may be used to determine dominant (or com-pelementary) wavelengths and purities on the CIE 1931 (x, y) and CIE 1964 (x_{10}, y_{10}) chromaticity diagrams, respectively. In each, three points are indicated that can serve as reference points (Table 7.6). The points for CIE ILL C and CIE ILL D_{65} may be used for the determination of domi-nant wavelengths (or complementary wavelengths) for the specification of colors of objects illuminated by daylight. In this case, the objects viewed are perceived as part of the environment. Point E is used when the color produced by light from a luminous object is being considered and also when an illuminated nonluminous object is viewed in surroundings that are sig-nificantly darker than the object. The chromaticity of the equal-energy light E is pertinent because it is considered neutral with respect to the dark field [Ref. 6.7, p. 846]. Because the values for λ_D (or λ_c) and for p_e depend on which reference point is used, a precise color specification must identify the illuminant.

A color specification in terms of dominant (or complementary) wave-length and purity is sometimes preferred to the standard $CIE(x, y, Y)$, be-cause it suggests immediately a perceived hue and saturation. If we are given the values for x and y, it is usually necessary to plot the point on the chromaticity diagram in order to gain some idea of color quality. Another advantage arises in comparing two colors that do not differ much. Compar-ison of their values for x and y can lead to a rough idea of their differences, but given values for their λ_D (or λ_c) and p_e we can tell relatively quickly how they differ in perceived hue and saturation [Ref. 6.2, p. 118].

Purity is only an approximate correlate of perceived saturation. Al-though purity and perceived saturation increase from the reference point outwards, they do not necessarily increase by similar steps. Furthermore, although purity, by definition, reaches 100 % at the spectrum locus and the purple line, saturation has been shown to attain only a maximum value that varies with dominant, or complementary wavelength [4.2, 4.3]. If the colors of two paint samples have the same purity but different dominant wavelengths, it is not unusual for the perceived saturations to be different [Ref. 1.20, p. 136]. Dominant (or complementary) wavelength, although a useful indicator of perceived hue, frequently does not accurately indicate equivalence or difference of hue. Along a straight line of constant dominant wavelength from the chromaticity point for the illuminant to the spectrum locus (or purple line), the perceived hue may vary significantly. Lines of constant perceived hue radiating from the reference point are, for the most part, not straight (Sect. 6.2).

Table 5.1 presents the CIE 1931 (x, y, Y) color designations, dominant wavelengths, and purities of a series of artists' pigments (CIE ILL C) (Sect. 5.2).

6.5 An Approximately Uniform CIE Chromaticity Diagram

The CIE colorimetric system was devised to provide a rigorous means of color specification by the principle of color matching with mixtures of three standard primaries. The system performs this function extremely well. It should not have been surprising, although it must have been disappointing, to have found, soon after the CIE 1931 (x,y) chromaticity diagram had come into use, new experimental data that revealed that, at constant levels of luminance or luminance factor, equal distances between pairs of points in different regions of the diagram usually did not correspond to equal differences of perceived color [Ref. 4.4, p. 129].

The nonuniformity of the diagram is immediately evident from the large region occupied by greens and the relatively tiny regions into which the reds and blues are crowded (Fig. 7.1). For example, in Fig. 6.14 points G_1 and G_2 represent the chromaticities of two greens of the same green hue. Their perceived color difference is equal to the perceived difference between two red purples R_1 and R_2 of identical red purple hue. The luminance factor for the four colors is the same, $Y = 0.20$. Although the perceived color differences are the same, the distance measured between points G_1 and G_2 is three times that between points R_1 and R_2.

From time to time attempts have been made to modify chromaticity diagrams or to produce others that would represent all color differences perceived as equal by equal distances between pairs of points. In 1960, the CIE provisionally recommended a transformation of the chromaticity diagram, for which the chromaticity coordinates were given by (u,v), in place of (x,y)

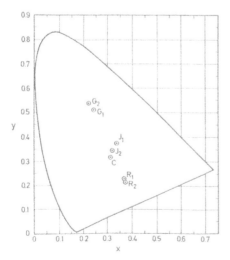

Fig. 6.14. CIE 1931 (x,y) chromaticity diagram showing the chromaticities of two pairs of colors of perceptually equal color difference $(G_1$ and G_2; R_1 and $R_2)$ at constant luminance factor $(Y = 0.20)$, and the chromaticity of the colors of a paint sample before (J_1) and after (J_2) fading with no observed change in luminance factor $(Y = 0.893)$. (Sect. 7.10)

of the 1931 diagram. While x, y, and z are the fractional amounts of the red, green, and blue primaries in a matching mixture used to specify the chromatic quality of a color (Sect. 6.2), u and v are counterparts of x and y empirically devised to produce a more uniform diagram. The quantities u and v may be calculated from the values of x and y [Note 6.2]. Examples of the CIE 1960 (u, v) chromaticity diagram are presented in [Refs. 4.4, Fig. 8.19; 5.8, p. 140].

The CIE 1960 (u, v) chromaticity diagram was used frequently for over a decade when uniformity of spacing was desired. Now the very similar CIE 1976 (u', v') chromaticity diagram is employed in its place (Fig. 6.15).

The modification in the 1976 version consists of one change: the parameter v' has been made equal to 1.5 times v [Note 6.2]. The modification was introduced at a time when the CIE also presented new recommendations for color-difference measurement. When CIELUV color differences are employed, the corresponding CIE 1976 (u', v') chromaticities are often also of interest (Sect. 7.10).

Like the (x, y) chromaticity diagram, the (u, v) and (u', v') diagrams possess both simplicity and the advantage of representing additive mixture by straight lines (Sect. 7.3). Other diagrams have been proposed with perhaps slightly better uniformity, but they do not possess the straight-line feature for additive mixture. The uniformity of the (u', v') diagram may be judged in Fig. 6.16 where curves of constant Munsell Chroma (8, 12, and 16) (at Munsell Value 5, $Y = 0.2$) are shown. The Munsell Chroma lines would be equally spaced concentric circles centered on the chromaticity point for CIE ILL C if the (u', v') diagram were perfectly uniform. (It is assumed that each curve of constant Munsell Chroma represents constant perceived saturation [Note 8.1].)

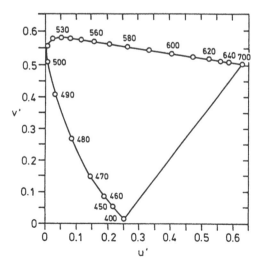

Fig. 6.15. CIE 1976 (u', v') chromaticity diagram. The wavelength [nm] of monochromatic light is indicated along the spectrum locus

An example of a (u'_{10}, v'_{10}) chromaticity diagram is shown in Fig. 6.17. The grid of points represents a set of perceptually equally spaced color samples (of equal lightness) discussed in Chap. 9. Here, the uniformity of the grid is an indication of the uniformity of the chromaticity diagram.

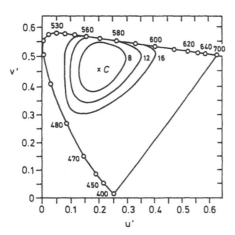

Fig. 6.16. CIE 1976 (u', v') chromaticity diagram. Curves are shown that pass through the chromaticity points of colors of constant Munsell Chroma (8, 12, and 16) and Munsell Value 5 (luminance factor $Y = 0.2$). Point C represents the chromaticity of CIE ILL C. (From [Ref. 4.3, Fig. 10])

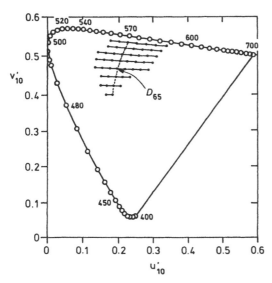

Fig. 6.17. CIE 1976 (u'_{10}, v'_{10}) chromaticity diagram. Chromaticity points are shown for perceptually equally spaced colors of a set of OSA-UCS chips at constant lightness level $L = 0$. The point for the chromaticity of CIE ILL D_{65} coincides with that for the neutral gray chip $(0, 0, 0)$ (Chap. 9). If the diagram were perfectly uniform, the points would describe a square grid. (From [Ref. 3.4, Fig. 6]; reproduced with the permission of John Wiley & Sons Inc., New York)

6.6 Metamerism and the CIE System

In his book on colorimetry, Wright states that "metamerism, which may be defined as the phenomenon of identity of colour appearance between stimuli of different spectral compositions, lies at the heart of colorimetry" [Ref. 6.2a, p. 138]. One illustration of this is found in Sect. 6.2 where the construction of a chromaticity diagram is demonstrated with the use of real stimuli (red R, green G, and blue B lights) serving as primaries. At any given luminance level, any point within the triangle shown in Fig. 6.1 or in Fig. 6.7 represents a metameric set. Included in any one metameric set is a mixture of stimuli G, R, and B which matches and hence identifies the single perceived color of all the metamers in the set. The color is specified by the relative amounts of G, R, and B required to match any one of the metamers. The fact that a mixture of three (and sometimes two) primaries can be used to match and thereby, by their relative amounts, be used to specify a color illustrates the key role of metamerism in colorimetry.

The same situation applies, of course, to the CIE system. The particular merit here, however, is that *all* visible stimuli are included, a feature made possible by the adoption of three imaginary primaries. The above principle of matching is embodied in one of the laws of color synthesis stated by the mathematician H.G. Grassmann (1809–1877) in 1853 [Refs. 1.20, p. 120; 1.39, p. 73].

Metamerism also plays a role in the mixing of lights (additive mixture, Sect. 7.3), which may be demonstrated with the use of Fig. 6.7. The straight dashed line CR represents the chromaticities of a gamut of colors obtained by mixing stimuli C (cyan) and R (red). Stimulus P produces pink, one of the colors in the gamut. But (in accordance with another of Grassmann's laws) the pink can be produced equally well by a mixture of metamers R' (which matches R) and C' (which matches C). The resulting P' (which matches P) has a different spectral power distribution (wavelength composition), but, at the luminance level considered, P and P' produce visually identical pinks. Stimulus P can be matched by mixtures of any two stimuli, at the same luminance level, whose chromaticity points fall on opposite sides of the point for P on any straight line drawn through that point. This straight-line mixture principle helps us to comprehend the large numbers of metamers that can be found in metamer sets [Ref. 2.5, p. 51].

It should be emphasized that what is a metameric set for one person with normal color vision is not necessarily exactly a metameric set for another person. Hence, a perfect match of two colors for one person may not correspond to the match for another one (Sect. 5.5). The metameric sets of the CIE(x, y, Y) system are those of the 1931 standard observer.

7. Diverse Applications of the CIE Chromaticity Diagram

7.1 Color Names for Lights

The CIE 1931 (x, y) chromaticity diagram is primarily a tool for those concerned with colorimetry and color specification. There are, however, a number of other applications to which it may be put and which are relevant in art and design.

First of all, the CIE chromaticity diagram can serve as a kind of *color-name map* for lights. K.L. Kelly has proposed the division of the diagram into color-name zones that form the map shown in Fig. 7.1 [4.1, 7.1]. The color names that he assigned are listed in Table 7.1. For the most part, the zones designate hue ranges. The color names do not reveal variations of purity, except by the inclusion of pinks, and do not vary when the luminance is changed (lights are not perceived as either black or gray).

In the large central oval area of Kelly's diagram, labelled U, no color names have been proposed. The hues of colors represented by the chromaticities in this zone vary in prominence from indefinite [Ref. 2.5, p. 52] to faint. The color produced by light from an incandescent-tungsten-filament lamp, typified by CIE ILL A (point A in Fig. 7.1), could be said to be a faint yellowish orange.

Kelly's diagram shows point C at the center (for CIE ILL C) from which the zone lines radiate. In Fig. 7.1, points have been added for illuminants CIE ILL D_{65} and CIE ILL B and for the equal-energy source E to show the locations of their chromaticities within an added sausage-shaped area [Ref. 2.5, p. 51] (dashed line) that could be designated as the achromatic or white zone (Sect. 7.4).

The large zone assigned to green G and the relatively small zone to red R does not mean that there are more greens than reds. If chromaticity points were plotted for colors of equal color difference and of equal luminance over the whole diagram, they would be found to be more densely spaced in the red zone than in the green zone (Fig. 6.14). This nonuniform spacing is considered a disadvantage inherent in the CIE (x, y) chromaticity diagram (Sect. 6.5).

On Kelly's diagram, the wavelength scale for monochromatic radiation is shown along the spectrum locus. In Fig. 7.1, this has been changed to show instead values of wavelength and complementary wavelength at

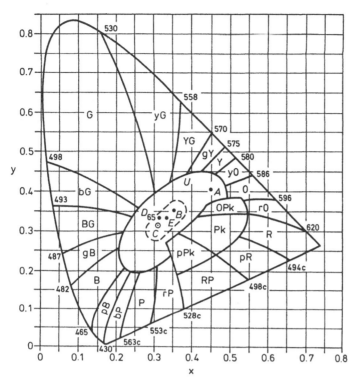

Fig. 7.1. Kelly's map for determining color names for lights (Table 7.1). The intersections of the hue boundaries with the spectrum locus and the purple line are indicated by wavelength and complementary wavelength, respectively. CIE 1931 (x, y) chromaticity diagram. (Modification of [Ref. 4.1, Fig. 1])

Table 7.1. Color names for lights (Fig. 7.1) [4.1]

pB	Purplish blue	O	Orange
B	Blue	OPk	Orange pink
gB	Greenish blue	rO	Reddish orange
BG	Blue green	Pk	Pink
bG	Bluish green	R	Red
G	Green	pR	Purplish red
yG	Yellowish green	pPk	Purplish pink
YG	Yellow green	RP	Red purple
gY	Greenish yellow	rP	Reddish purple
Y	Yellow	P	Purple
yO	Yellowish orange	bP	Bluish Purple

points where hue-boundary lines intersect the spectrum locus and purple line. These values are used to designate the perceived hue ranges in the spectrum given in Table 4.2.

Kelly's color-name zones are also useful as a quick and approximate way of identifying the colors of objects from their chromaticities (x, y), for example, from CIE 1931 (x, y, Y) in a color specification. However, it should be noted that because Kelly's nomenclature is intended for lights (for which dark and grayish colors are not perceived) it does not include such color names as olive green and brown. For example, at a chromaticity of (0.540, 0.410) the color of an object is deep orange at a luminance factor $Y = 0.20$; strong brown at $Y = 0.12$; and deep brown at $Y = 0.03$ [7.2]. For colored light of the same chromaticity, however, the color would be orange, regardless of the luminance. A more satisfactory method for designating names for the colors of materials, the ISCC-NBS method [7.2], is discussed in Sect. 10.1.

7.2 Additive Complementary Color Pairs

The CIE chromaticity diagram [either 1931 (x, y) or 1976 (u', v')] can serve as a kind of color circle for identifying *additive complementary color pairs.* This is not surprising; we need only recall the diametric placements of complementary colors in the Maxwell triangle and in the six-member color circle (Fig. 6.6). The notion of additive complementary colors (Sect. 6.2) can be illustrated by *additive color mixture.* If two beams of light of widely different hue can be adjusted in intensity so that their mixture will produce a white disk on a white wall, the original colors are said to be complementary. Likewise, if the colors of two different papers are mixed, such as when sectors of colored paper are viewed on a rapidly rotating disk (*color mixture by averaging*), and if a neutral gray can be produced by adjusting the areas of the sectors of the two colors, the two colors are said to be complementary.

A straight *mixture line,* which connects the two points that represent the chromaticities of the two colors on a chromaticity diagram, is the path on which all points fall that represent the chromaticities of all possible additive mixtures of the two colors (either additive color mixture or mixture by averaging) (Sect. 6.2). If the straight mixture line passes through the central achromatic region (the vaguely defined region indicated by the dashed oval in Fig. 7.1), then an achromatic mixture is possible. Two colors are said to be additive complementary pairs if a straight line drawn between their chromaticity points passes through the chromaticity point of the hueless illuminant. Thus additive complementary pairs can be determined precisely with respect to a selected reference white, for example point E (equal-energy light source) in the case of colored light, or point C (CIE ILL C) or point D_{65} (CIE ILL D_{65}) in the case of objects illuminated by daylight.

The chromaticities of the colors of two monochromatic beams M (494 nm) and N (640 nm) are indicated in Fig. 7.2. Their mixture line passes through point E, showing that the colors are complementary with respect to the equal-energy source E. The dashed line that connects points Q and K shows that the colors Q and K, a blue and a yellow, are complementary

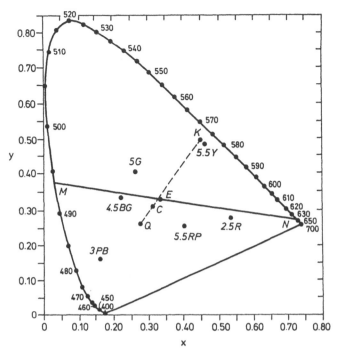

Fig. 7.2. CIE 1931 (x, y) chromaticity diagram used in the determination of complementary color pairs (see also Figs. 6.12,13). The opposing pairs of colors in a six-member color circle (Plate I, *upper circle*) are shown here to be the near-complementary color pairs (additive): 3PB and 5.5Y; 4.5BG and 2.5R; 5G and 5.5RP

with respect to CIE ILL C. These are also the colors of two paint samples discussed in Sects. 7.8 and 9. In color mixture by averaging, such as for color samples on a rapidly rotating disk, the dashed line indicates the path of the chromaticities of the colors of their mixtures. But, as is shown later, the points that represent various *subtractive mixtures* of the same two paints would fall on a curve that would not pass through the achromatic zone (Fig. 7.14). No mixture of the two paints can produce neutral gray. The question of whether the colors of materials are complementary is answered by whether the straight line (for color mixture by averaging) passes through or close to the reference point.

The chromaticities of the three pairs of colors employed in the upper six-member color circle in Plate I are identified by six points in Fig. 7.2. The pairs are 3 PB, strong blue (178), and 5.5 Y, vivid yellow (82); 4.5 BG, brilliant bluish green (cyan) (159), and 2.5 R, vivid red (11); and 5 G, strong green (141), and 5.5 RP, strong purplish red (magenta) (255) (ISCC-NBS color names and centroid numbers, Chap. 10 [7.2]). When straight lines are drawn to connect these pairs of points, they are found to pass rather close

69

to point C. Hence the pairs are approximately complementary with respect to CIE ILL C.

In the above discussion, additive complementary color pairs are taken to be those that can produce white or neutral gray by additive color mixture or by color mixture by averaging. This is the psychophysical concept of complementary colors. In psychology, the word "complementary" is often used somewhat differently, for example to describe the colors that are perceived in two visual phenomena, afterimages and simultaneous contrast (which includes colored shadows) [Ref. 2.5, p. 222]. To differentiate between the psychophysical and psychological concepts, the terms "additive complementary color pairs" and "afterimage complementary color pairs" (Sect. 11.7) are used here, following the example of [5.21].

7.3 Colors Obtainable by Mixing Light

The CIE 1931 (x, y) chromaticity diagram can serve in a way that should be of interest to those who work with colored lights, lasers, and phosphors (for example, color-television phosphors) as art media. It can be used to predict the chromaticity of colors obtainable by mixing two or more light beams of different color [Ref. 7.3, p. 139]. (The CIE chromaticity diagram is sometimes called a *mixture diagram* [Ref. 6.7, p. 842].) If, for example, purplish blue light Q is mixed with red light R by combining the beams from two projectors, the chromaticity of the resulting color is located at some point on the straight *mixture line* that connects the chromaticity points Q and R (Fig. 7.3).

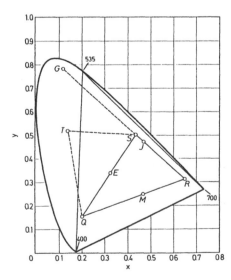

Fig. 7.3. Chromaticity gamuts available for mixtures of several beams of colored light

70

The precise location on the mixture line of the chromaticity point M for the color of the mixture depends on the relative amounts of Q and R. For color-mixture calculations, the amount of each of the two beams is given by the measured luminance Y divided by the chromaticity coefficient y [Refs. 2.1, p. 236; 7.4]. If, for beam Q, $Y = 30$ and $y = 0.15$ (see Fig. 7.3), and, for beam R, $Y = 90$ and $y = 0.30$, then the amount of Q is 30/0.15, which is 200, and that of R is similarly 90/0.30, which is 300. The amount of mixture is $200 + 300$, i.e., 500, and the fractional amount of R in the mixture is 300/500, which is 0.60. The mixture is predominantly red: the mixture point M is located on the mixture line 0.60 (or 60 %) of the way from Q to R. {The amount of Q (or R) is the sum of its tristimulus values: $X + Y + Z$ [Note 6.1].}

As in stage-lighting practice, after an estimation of point M is made to represent the chromaticity of the color desired, it is necessary to find two light filters that provide beams, say Q and R, that when mixed in the required proportions will produce M. The needed proportions can be found by adjusting the intensities of the two beams [Ref. 7.3, p. 141].

If the red light is mixed with the greenish yellow light represented by point S (Fig. 7.3), the chromaticities of the gamut of colors available are represented by the mixture line that connects R and S. The extension of the mixture line up to G in the green region of the diagram demonstrates that yellow J can be produced by an additive mixture of red R and green G (Sect. 5.6).

The mixture line between Q and S applies to the gamut of mixtures obtainable by mixing purplish blue and greenish yellow light. Because the line passes through point E, the two colors are additive complementary pairs with respect to the equal-energy source E. In proper proportions, they add to produce hueless (white) light at E.

Now let us turn our attention to the gamut of chromaticities of colors obtainable by additive mixture of three or more beams of light of different colors. The gamut available with three beams of different hues is given by a triangle. In Fig. 7.3, one such gamut is indicated by the triangular area formed by joining points Q, R, and S. The gamut can be increased by extending the area to reach other points outside the triangle. For example, if point T is added, representing a fourth beam of another hue, the new gamut is given by the area bounded by four lines formed by joining successively points Q, R, S, T, and Q.

The large gamut that may be produced by mixtures of three monochromatic light beams (the Hardy-Wurzburg triangle), to which reference is made in Sect. 5.9, is illustrated by the triangle formed by connecting the points on the spectrum locus labelled 700, 535, and 400 nm [Ref. 2.1, p. 238]. Because monochromatic light can be produced by lasers or can be isolated readily from laser beams consisting of light of several wavelengths, lasers offer a rich potential source of large gamuts of colors. Unfortunately, however, costly equipment may be required to produce such mixtures, and provisions may have to be made to eliminate serious safety hazards.

It should be recalled that the CIE 1931 (x, y) chromaticity diagram is considered suitable especially for angles of vision between $1°$ and $4°$ (Sect. 6.3). Those interested in artistic applications of colored lights where large angles of vision apply should consider using the CIE 1964 (x_{10}, y_{10}) chromaticity diagram (Fig. 6.13) if accuracy is of importance.

In the above discussion, reference is made only to colors perceived by the neutral-adapted eye. In a theater our eyes are commonly adapted to stage lighting produced by incandescent-tungsten-filament lamps (approximated by CIE ILL A). Under the latter conditions, lamplight is seen as white, and the perceived colors of objects differ somewhat from those in daylight (Sect. 11.4) [7.3].

7.4 Light Called "White Light"

Light that produces an achromatic (hueless) visual response is commonly called "white light". Measurements show that light that produces such a response is not characterized by a unique chromaticity, but rather by a vaguely defined gamut of chromaticities suggested by a sausage-shaped area whose approximate length is indicated by the points for 4000 and 10 000 K on the color-temperature curve (Fig. 7.21) [Ref. 2.5, p. 51]. The dashed line (small oval) in Fig. 7.1 is intended to outline roughly the sausage-shaped area. Light represented by any point within that area (E and CIE ILL B, C, and D$_{65}$) evokes an achromatic response; it is white. By comparison, light from an ordinary incandescent (tungsten-filament) lamp (typified by CIE ILL A) does not; it is a faint yellowish orange. But with adaptation, incandescent-lamp illumination appears white (Sect. 11.4).

Because sunlight can be dispersed into a spectrum of pratically all wavelengths in the range from 380 to 780 nm, it is often erroneously assumed that white light is necessarily a mixture of light of all wavelengths in that range. But is should be remembered that white light of a given chromaticity can be produced also by many mixtures whose wavelength composition does not include all the wavelengths. An extreme example is provided by a mixture whose wavelength composition is given by two wavelengths, such as the two complementary monochromatic beams M and N indicated in Fig. 7.2.

It is easy to demonstrate how three beams of monochromatic light, H (490 nm, blue green), K (570 nm, greenish yellow) and L (620 nm, red) (Fig. 7.4), can be combined to produce a beam of white light (represented by E). The demonstration may be begun by combining beams H and K to produce the mixture J. Then beam L is added to J to produce the final mixture E. The resulting white light is composed of only three wavelengths. A similar demonstration could be made for the production of white light from

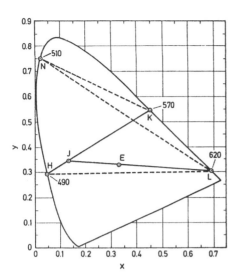

Fig. 7.4. Production of white light from three beams of colored light

four, five, or any number of different, appropriately selected monochromatic beams.

On the basis of the discussion in Sect. 7.3, beams H, K, and L can be expected to provide a combination that forms a white mixture E, because the triangle formed by connecting the three points *encloses* E. On the other hand, it is evident that monochromatic beams K, L, and N cannot produce a white mixture (Fig. 7.4); the chromaticity gamut defined by the triangle KLN does not contain any point for white.

The fact that mixtures of light, such as one of H, K, and L, match daylight precisely does not mean that the perceived color of an object illuminated by daylight and by each of its matching mixtures will be the same. Generally, the color will be different under each illumination. Only in the case of a white object, which reflects almost all of the light it receives, is it certain that the color (white) will not change (this is the condition under which a match is made). A vase that is green in daylight is not green when illuminated by the white mixture of beams H, K, and L, because no light is provided in the wavelength region from 500 to 560 nm for scattering to the eye. Light mixtures of equal luminance that match (metameric illuminants) may be of potential interest in art (Sects. 5.5 and 7.13). In the above illustration, a mixture of three monochromatic beams was employed, but, generally, there is no reason why the beams of light should be monochromatic.

In general, however, people prefer lighting in which objects (especially their faces) appear in their "natural" coloring. This subject is dealt with under the designation *color rendering* in the domain of illumination engineering (Sect. 7.14).

7.5 The Color Limits for Materials (Paints, Inks, Dyes, etc.)

The CIE chromaticity diagram can be used to present a maximum-gamut map that defines the ultimate limits of the gamuts of all colors produced by colorants. Of course, practical limits, imposed by the availability and cost of pigments, dyes, and light sources, restrict what an artist or designer can employ, but these limits are pushed back as technology advances toward the limits of what is possible.

The ultimate limits of the gamut of all colors produced by *light from luminous sources* are set by the tongue-shaped spectrum locus and the straight purple line on the chromaticity diagram. These limits are not modified by the luminance Y under normal conditions for perceiving colors.

It was pointed out in Sect. 6.3 that the CIE chromaticity diagram applies not only to the colors produced by light coming directly from luminous sources but also to the colors of objects, because color measurement is performed on the light received after it has been scattered by or transmitted through the objects. The important difference is that *only certain regions* of the tongue-shaped area are available to represent the chromaticities of the colors of scattered or transmitted light − light that remains after selective absorption occurs in nonluminous, nonfluorescent objects. In these cases, the shapes and sizes of the regions are related to the luminance factor Y, for opaque objects (or luminous transmittance Y, for transparent objects). The limits of the regions have been determined precisely for CIE ILL A and CIE ILL C by the psychophysicist D.L. MacAdam and are known in the United States as the *MacAdam limits* [Refs. 4.4, p. 122; 7.4,5]. The German mineralogist S. Rösch (1899–1984) reported similar work on the subject somewhat earlier (1929), and in German color literature his name is associated with the same concept [Refs. 3.14, p. 341; 5.3, Fig. 14.09(2)]. The limits are sometimes called the *pigment limits,* but they apply to dyes equally well.

The condition for which the luminance factor is taken equal to 1.0 ($Y = 1.0$) represents whiteness equal to that of the hypothetical *100 %-white* surface. It represents total scattering of incident light without selective absorption and, hence, without color change. This is approached by snow and finely ground table salt, for example. Similarly, transparent glass would be *colorless* ($Y = 1.0$) if light of all wavelengths passed through it without appreciable reduction of intensity or change of wavelength composition. Thus for a situation in which $Y = 1.0$, the chromaticity of the color of the scattered or transmitted light is identical to the chromaticity of the color of the incident light. The discussion below concerns the MacAdam limits for the colors of *nonfluorescent opaque materials in daylight illumination* (CIE ILL C), in which the only color that can be produced at $Y = 1.0$ is white. This condition is represented by point C on the chromaticity diagram, the location of the chromaticity for CIE ILL C (Fig. 6.12). (The discussion here also applies to *nonfluorescent transparent materials* for which Y is the luminous transmittance.)

A luminance factor (or luminous transmittance) Y that is less than 1.0 signifies that part of the light received by the object is absorbed; the remainder is scattered (or transmitted) and reflected. At chromaticity point C and over the range of Y from 1.0 to zero, there is the achromatic gamut varying from white (or colorless) through the neutral grays to black. Thus, at $Y = 0.60$, the color is light gray; at $Y = 0.25$, medium gray; at $Y = 0.10$, dark gray; and below $Y = 0.05$, essentially black [7.2]. It should be noted that a luminance factor of 0.60, for example, implies 40% absorption of psychophysical light, *not* 40% absorption of received visible radiant energy (Sect. 6.1).

The previous paragraph concerns neutral grays at various levels of Y, when an object is illuminated by a source typified by CIE ILL C. The next step is to consider the limits of chromatic colors possible at the various levels of luminance factor (or luminous transmittance) Y. For a fixed value of Y and the one illuminant, the chromaticities of the colors are confined to a region of the chromaticity diagram bounded by a MacAdam limit.

At $Y = 1.0$, the region of the chromaticity diagram available for the colors of objects is restricted to a point, point C – that is, to one color only: white (or colorless). [Note that by giving the chromaticity, the luminance factor (or luminous transmittance), and the illuminant CIE ILL C, we specify the color of an object.]

The small region on the chromaticity diagram defined by the MacAdam limit at $Y = 0.95$ is shown in Fig. 7.5. Also indicated are three color zones (dashed lines) based on the color-name system [7.2] described in Sect. 10.1. The point shown represents the chromaticity of a light neutral gray. Outside the MacAdam limit, no colors with $Y = 0.95$ are possible for *nonfluorescent* objects.

When the luminance factor (or luminous transmittance) Y is decreased, the ranges of possible chromaticities described by the MacAdam limits, expand. Figure. 7.6 shows the MacAdam limits at 11 levels of Y, from $Y = 1.00$ (a point) down to $Y = 0$. At $Y = 0.90$, the limited region is approximately rectangular, as it is at $Y = 0.95$ (Fig. 7.5). At $Y = 0.95$, the gamut is rather small, being limited primarily to the yellow greens from pale to brilliant and vivid. At $Y = 0.90$, the gamut includes yellows and more yellow greens. Beginning at about $Y = 0.70$, the gamuts crowd the spectrum locus in the red, orange, and yellow regions. At $Y = 0.10$, the limit is relatively large; at that level some colors are seen to be dark [7.2].

At $Y = 0$, the MacAdam limit coincides with the spectrum locus and purple line of the chromaticity diagram, and the colors (for all hues) are of maximum darkness: black [Note 7.1] [7.6]. A sketch of a model of the color space (Fig. 7.6) is given in Fig. 8.2; a stereoscopic pair of photographs of a model is given in [Ref. 4.4, Fig. 7.25].

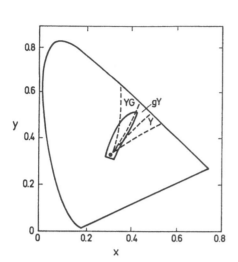

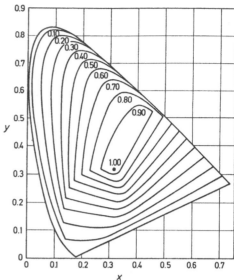

Fig. 7.5. Chromaticity limit (MacAdam limit) for colors of nonfluorescent materials, luminance factor (or luminous transmittance) $Y = 0.95$ (CIE ILL C) [7.5]. CIE 1931 (x, y) chromaticity diagram. ISCC-NBS color-name zones: Y (yellow), gY (greenish yellow), and YG (yellow green) [7.2]

Fig. 7.6. Chromaticity limits (MacAdam limits) for colors of nonfluorescent materials. Luminance-factor (or luminous-transmittance) range from $Y = 0$ to $Y = 1.00$ (CIE ILL C). CIE 1931 (x, y) chromaticity diagram. (Based on [Refs. 4.4, Fig. 7.23; 7.5])

Plate II shows a number of glossy samples that have luminance factors equal to about $Y = 0.30$. The color samples are intended to provide a varied display at one level of luminance factor and to show the distribution of colors on the chromaticity diagram. Some of the samples (1–10) were cut from specimens in a commercial pigment catalog [7.7], and others (11–22) were cut from standard Munsell color chips (Sect. 8.4). Data relevant to the samples are given in Table 7.2. Faithful reproduction of the colors (originally accurately represented by their designations) cannot be assured in Plate II.

The MacAdam limits apply to colors that are not diluted by surface-reflected light (Sect. 5.2). Calculations of limits have been published that take into account surface reflection of 4 % of a nearly perpendicularly incident light beam [Refs. 5.8, p. 133; 7.8]. If the incident light is diffuse, yet nearly perpendicular, 9.2 % is reflected at the surface [3.2]. The resulting limits are decreased significantly; an effect that should be remembered when matt paint colors are being considered.

Figure 7.7 shows the chromaticity points for the colors of samples of artists' acrylic paints based on data published by one manufacturer. Not

Table 7.2. Colors of samples shown in Plates II and VII

Sample number	ISCC-NBS color name and centroid number	Luminance factor, Y	Munsell notation Hue Value/Chroma
1	Strong bluish green (160)	0.342	4.5BG 6.34/8.9
2	Brilliant blue (177)	0.301	3.0PB 6.00/10.0
3	Light purplish blue (199)	0.332	5.0PB 6.26/6.8
4	Light purple (222)	0.320	6.0P 6.16/6.4
5	Light reddish purple (240)	0.251	0.4RP 5.54/8.6
6	Deep purplish pink (248)	0.327	4.3RP 6.22/11.5
7	Deep pink (3)	0.329	5.2R 6.24/7.2
8	Moderate reddish orange (37)	0.299	7.0R 5.99/10.2
9	Deep pink (3)	0.280	2.5R 5.82/13.9
10	Strong reddish orange (35)	0.278	8.5R 5.80/12.6
11	Strong orange (50)	0.301	5.0YR 6.00/10.0
12	Vivid orange (48)	0.301	2.5YR 6.00/16.0
13	Deep yellow (85)	0.301	5.0Y 6.00/10.0
14	Dark yellow (88)	0.301	5.0Y 6.00/6.0
15	Dark grayish yellow (91)	0.301	5.0Y 6.00/4.0
16	Light grayish olive (109)	0.301	5.0Y 6.00/2.0
17	Greenish gray (155)	0.301	5.0GY 6.00/1.0
18	Strong yellow green (117)	0.301	5.0GY 6.00/10.0
19	Vivid yellowish green (129)	0.301	10.0GY 6.00/12.0
20	Brilliant green (140)	0.301	5.0G 6.00/10.0
21	Brilliant greenish blue (168)	0.301	2.5B 6.00/8.0
22	Pale blue (185)	0.301	2.5B 6.00/2.0

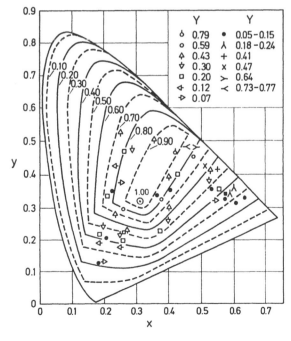

	Y		Y
◊	0.79	●	0.05–0.15
○	0.59	λ	0.18–0.24
△	0.43	+	0.41
▽	0.30	×	0.47
□	0.20	⊳	0.64
◁	0.12	≺	0.73–0.77
▷	0.07		

Fig. 7.7. Chromaticities of the colors of artists' acrylic paints (nonfluorescent) in relation to the chromaticity limits (Mac Adam limits). At Y = 1.00 (white) the limit is represented by one point (for CIE ILL C). CIE 1931 (x, y) chromaticity diagram. Points indicated by circles and triangles, [7.9]; other points, [7.10]

shown are points for very dark paints for which the luminance factor Y is below 0.05. The plot is interesting because it demonstrates the general distribution of the hues. It also provides an example of the application of the MacAdam limits. The position of a chromaticity point for the color of a paint can be compared with its MacAdam limit at the same level of luminance factor. Paints that have luminance factors of about 0.20 (points shown as squares), for example, are compared on the chromaticity diagram with the MacAdam limit for $Y = 0.20$. Because these particular paints are known to contain pigments of good permanency, the proximity of their chromaticity points to the limit gives some indication of the possible improvements of color purity that might be hoped for in new stable pigments. Of course, allowance should be made for the fact that water-based acrylic paints produce matt films and that, if they were made glossy by the addition of a layer of clear acrylic lacquer, their color purities would increase and the chromaticity points for the colors would be closer to the MacAdam limits [Sect. 5.2].

Figure 7.7 could be misleading. We must not interpret the expanding MacAdam limits of the chromaticities of possible colors accompanying decreasing luminance factor Y as an indication of an increasing number of colors. Such an interpretation would lead to the conclusion that at $Y = 0.1$ the number of colors is relatively large and at $Y = 0$ the number of colors is largest, when, in fact, only one color is possible, black. A MacAdam limit drawn on a CIE 1931 (x, y) or CIE 1964 (x_{10}, y_{10}) chromaticity diagram shows the theoretical boundary of the chromaticities of nonfluorscent colors of objects at some specific level of luminance factor; a limit on those diagrams does not reveal the relative number of possible colors, considered, say, on the basis of perceptually uniform spacing. It is true that starting at $Y = 1.00$, where there is but one color (white), the number of equally spaced colors increases with decreasing Y, but what Fig. 7.7 does not show is that at a luminance factor of about $Y = 0.2$ [Note 7.1] the number of possible equally spaced colors reaches a maximum and that, thereafter, as the luminance factor is decreased, the number of colors decreases, finally decreasing to one color, black, at $Y = 0$.

The possibility of misinterpretation is avoided in some other color systems (Sects. 8.2, 4–7) in which black is represented by one point. But in those systems, samples are generally displayed (sometimes with the MacAdam limits shown) only in single arrays (for example, an array at constant Munsell Value, as in Fig. 8.10 and Plate VII), whereas, on a CIE(x, y) chromaticity diagram a series of arrays of samples at different levels of luminance factor is conveniently accommodated because there the MacAdam limits fan out in clearly distinguishable steps as Y is decreased (Fig. 7.6).

7.6 Fluorescent Paints and Dyes

The MacAdam limits (Sect. 7.5) apply only to the colors of nonfluorescent objects. The chromaticity points for the colors of fluorescent paints are commonly located outside the MacAdam limits, but they are never located outside the boundaries of the tongue-shaped chromaticity diagram — that is, in the territory of imaginary colors (Sect. 6.3, Fig. 6.9). An example of a point that falls outside the MacAdam limit for $Y = 0.55$ is specified by CIE(0.640, 0.355, 0.553), CIE ILL C [5.14]; it is given by point A in Fig. 7.8. The color of a nonfluorescent paint, used as a standard safety color, having about the same chromaticity (the same point A) is appreciably darker ($Y = 0.15$). It is well within the MacAdam limit shown for $Y = 0.15$.

A number of fluorescent dyes are commercially available. As mentioned earlier (Sect. 5.4), products sold as fluorescent pigments are generally dyes dissolved in a plastic base that has been hardened and ground to a powder. Unfortunately, presently available fluorescent materials have inferior light-fastness. However, by use of ample amounts in paints and by restricting exposure to light, the life of fluorescent paintings can be extended appreciably.

It is hoped that fluorescent colorants of great variety will eventually be developed. Evans mentioned the interesting possibility of extending the range of Munsell samples beyond the MacAdam limits by use of fluorescent pigments [3.9]. His studies of perception showed that the nonfluorescent region passes continuously into the fluorescent region. Fluorescent colors that are not fluorent might be used to fill in some gaps.

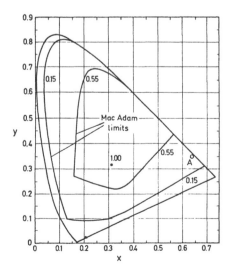

Fig. 7.8. Example of the chromaticity (A) of the color of a fluorescent paint that is located outside the chromaticity limit (MacAdam limit) at luminance factor $Y = 0.55$ (CIE ILL C). CIE 1931 (x, y) chromaticity diagram [5.14]

7.7 Iridescent Colors: Liquid Crystals

Iridescent colors are familiar to all of us in everyday life. They can be seen in nature often in the feathers of birds, in the wings of butterflies, in the scales of fish and beetles, and in sea shells [Refs. 7.11; 7.12, p. 56]. Sometimes the colors perceived are of high saturation, as in peacock feathers and certain beetle scales. We see iridescent colors in soap bubbles, in oil slicks on wet pavements, and in paints and plastics containing man-made pearlescent pigments [7.13]. Commonly we see today toys, jewelry, etc. decorated with a plastic film selectively embossed with diffraction gratings which, without colorants, produce patterns of vibrant spectral colors [7.14]. Iridescence can also be observed in many organic chemical products known as *liquid crystals*. One type, cholesteric liquid crystals, has recently become of interest as an art medium [7.15,16]. Working independently in the early 1970s, artist Yves Charnay in France [7.17] and physicist-artist David Makow in Canada [7.18] introduced the use of liquid crystals in their paintings.

Iridescence is produced in liquid crystals by the *selective reflection* of light. The physical mechanism occurring is called constructive interference. In effect, light of a narrow band of wavelengths from one half of the incident light is reflected from numerous layers in a liquid-crystal structure. Light of other wavelengths of that half of the incident light passes through the liquid-crystal film. All the other half of the incident light (typified by waves that are circularly polarized in the opposite manner) does not interact with the structure; it passes through the film without reflection. Unlike ordinary dyes and pigments, liquid crystals absorb practically no light (less than 1 %). In order to obtain a high-purity green color, for example, the liquid crystals capable of producing the hue should be painted on a black background that will absorb all the transmitted light, allowing only the selectively reflected light (green) to be seen [7.16].

If a liquid crystal is painted on a flat, white surface and daylight illumination (or another common white-light source) is employed, little or no color is produced because reconstituted daylight resulting from the additive mixture of the selectively reflected light *and* the light transmitted to and reflected from the white surface is seen by the eyes. When a colored background is employed, the resulting color is determined primarily by the additive mixture of the light scattered by the background and the light selectively reflected by the liquid-crystal film.

Of particular interest is what happens when one liquid-crystal film is applied over a film of another, on a black background. If the two liquid crystals have widely different wavelength bands of selective reflection, incident daylight is first subjected to selective reflection in the top film, and then the light that passes through to the second film is subjected once more to selective reflection. The light transmitted through the two films is absorbed by the black pigment of the background. The two wavelength bands of light reflected from the two films to the eyes produce the resulting color by additive mixture. This is in marked contrast to what occurs (subtractive mixture) when a beam of light is *selectively absorbed* as it passes through two light filters containing different dyes or pigments.

Figure 7.9 shows the spectral reflectance curve for two superimposed films of liquid crystals having widely different wavelength bands of selective reflection. Viewed individually, one film is seen as green, the other as red. The two superimposed films in daylight produce an orange red by additive mixture [7.18]. Figure 7.10 presents the spectral reflectance curve for another pair of superimposed films. Here the wavelength bands are much closer together. The dashed lines are the spectral reflectance curves for the two liquid crystals measured individually.

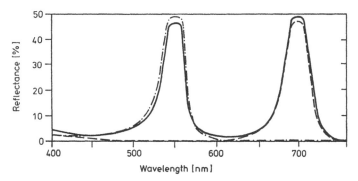

Fig. 7.9. Spectral reflectance curves for a pair of superimposed films of cholesteric liquid-crystal samples having widely different wavelength bands of selective reflection. ——: curve for the pair of films; - - -: curves for the individual films. (Modified from [Ref. 7.16, Fig. 3a]; reproduced with the permission of John Wiley & Sons Inc., New York)

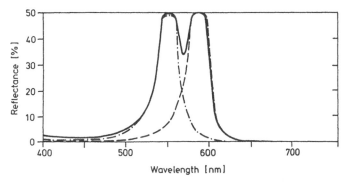

Fig. 7.10. Spectral reflectance curves for a pair of superimposed films of cholesteric liquid-crystal samples having moderately different wavelength bands of selective reflection. ——: curve for the pair of films; - - -: curves for the individual films. (Modified from [Ref. 7.16, Fig. 3b]; reproduced with the permission of John Wiley Sons Inc., New York)

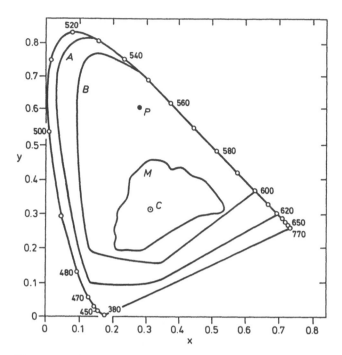

Fig. 7.11. Chromaticity (*P*) of the color produced by a cholesteric liquid-crystal film (Fig. 7.9, --- curve at left) on a black background in daylight (*C*) (CIE ILL C). Luminance factor $Y = 0.165$. The chromaticity limit *M* of the colors of the painted chips of the *Munsell Book of Color* is shown for comparison [7.19]. *Curve B:* theoretical limit ($Y = 0.165$) for the chromaticities of the colors produced by cholesteric liquid crystals; *curve A:* MacAdam limit ($Y = 0.165$) for ordinary nonfluorescent colorants. CIE 1931 (*x*, *y*) chromaticity diagram. (Modified from [Ref. 7.16, Fig. 5]; reproduced with the permission of John Wiley & Sons Inc., New York)

Spectral reflectance curves obtained for cholesteric liquid crystals do not exceed 50%. This, the theoretical maximum, can be explained by the fact that 50% of the incident light (of all wavelengths) is not available for selective reflection; it passes through liquid crystals unaffected [7.16] [Note 7.2]. The remaining 50% is available for selective reflection. In Fig. 7.11 we turn our attention to the chromaticity *P* of a green color produced by a liquid-crystal film at luminance factor $Y = 0.165$ (CIE ILL C). The purity of the color is much higher than that obtained with ordinary nonfluorescent pigments. That the purity is rather high is evident from the position of *P* relative to the range of chromaticities *M* of the colors of the collection of chips in the *Munsell Book of Color* [7.19]. Yet higher purities are possible, as is evident from the location of *P* in relation to the limit curve *B* ($Y = 0.165$) applying to this type of liquid crystal [Note 7.2]. Curve *A* represents the MacAdam limit (at $Y = 0.165$) that applies to the usual nonfluorescent colorants (Sect. 7.5) and paints, such as those used in making the Munsell color chips.

A painting in liquid crystals in the form employed by Charnay must be protected to prevent their movement, prolonged contact with oxygen of the air, and the collection of dust by a continuously wet surface [7.17][1]. Mylar film (polyethylene tetraphthalate) has been used successfully for covering painted surfaces and separating superposed films of different liquid crystals [7.17]. Some protection may be offered by the use of encapsulated liquid crystals – that is, minute droplets of liquid crystals enveloped in tiny spheres of gelatin or gum arabic. But much of the vividness characteristic of liquid crystals is diminished by encapsulation because it causes a reduction in the uniformity of the structural alignment in the liquid crystals [7.18].

Many liquid crystals exhibit a temperature dependence of their colors. Temperature affects, for example, the spacing of the reflecting layers in liquid crystals [7.18]. The result is that varying colors may be produced over a specific temperature range (say, 20°–30°C); outside this temperature range, the liquid-crystal film has a colorless appearance. Charnay has created individual paintings using different liquid crystals that produce colors in different temperature ranges [7.17]. In Plate III (color plate), one of his paintings that consists of different liquid crystals is shown at two temperatures, 18°C and 23°C, to illustrate a typical change in appearance.

When we view a liquid-crystal painting, the colors that we see change if we change our angle of view to the painting. This is explained by the fact that the specific selectively reflecting planes that affect what we see change when the angle of view changes [7.18]. The more oblique the angle of viewing, the more the hue is shifted toward greens or blues. Makow has utilized this property in painting certain of his sculptures and reliefs with encapsulated liquid crystals. Rounded surfaces and inclined planes acquire accents in different hues owing to different angles presented to the viewer. When a white form is painted with liquid crystals, a tint provided by the reflection hue will appear in the highlight and the remainder of the surface will show a tint of the hue of the transmitted light. This behavior is observed in pearls, which display iridescence, often showing bluish green in the highlight and rose tints on the remaining surface [Ref. 7.13, p. 379].

7.8 Mixing Paints

The CIE chromaticity diagram provides a convenient way of predicting the chromaticities of the colors of mixtures of light (Sect. 7.3). The fact that the mixture lines are straight (additive color mixture), and require only

---

[1] It has just come to my attention that these problems can be avoided by the use of a newly developed polymer liquid crystal of the polysiloxane type [D. Makow: Reflection and Transmission of Polymer Liquid-Crystal Coatings and Their Application to Decorative Arts and Stained Glass. Color Res. Appl. **8**, (3), 205–208 (1986)].

two points to establish them, accounts for the ease of the method. In the case of mixing pigments or dyes (subtractive color mixture), however, the mixture lines are very frequently curved and three or more points must be provided to establish them on the chromaticity diagram. Plate IV shows seven sets of four chromaticity points that represent a white pigment (titanium white), a colored pigment, and two mixtures of the colored and white pigments. The mixture line (sometimes called a colorant trace) for each series terminates at point C (CIE ILL C), which represents the chromaticity of the white. The color samples shown were cut from glossy specimens in a commercial pigment catalog [7.7]. The identification of each sample and relevant information are presented in Table 7.3.

The mixture lines in Plate IV reflect the changes of perceived hue (roughly represented by the dominant, or complementary, wavelength) and

Table 7.3. Identification of pigment mixtures and colors of glossy samples reproduced in Plate IV

Sample	C/W Pigment weight ratio Colored (C) White (W)[b]	ISCC-NBS color name	Luminance factor Y	Munsell notation Hue Value/Chroma
[a]1-A	100/0	Strong greenish yellow	0.503	9.0Y 7.47/10.5
1-B	50/50	Light greenish yellow	0.702	0.5GY 8.59/7.7
1-C	10/90	Pale greenish yellow	0.824	2.0GY 9.17/3.8
2-A	100/0	Moderate yellow	0.560	1.5Y 7.82/16.8
2-B	33/67	Brilliant yellow	0.652	2.5Y 8.33/10.9
2-C	5/95	Pale yellow	0.758	4.0Y 8.87/4.9
3-A	100/0	Vivid reddish orange	0.169	10.0R 4.67/16.4
3-B	33/67	Strong reddish orange	0.278	8.5R 5.80/12.6
3-C	5/95	Strong yellowish pink	0.500	7.5R 7.46/7.2
4-A	100/0	Deep reddish brown	0.043	10.0R 2.41/11.8
4-B	33/67	Strong red	0.103	2.5R 3.72/12.8
4-C	10/90	Strong purplish red	0.197	9.0RP 4.99/12.0
5-A	100/0	Very deep red	0.0138	8.2R 1.12/7.6
5-B	25/75	Deep purplish red	0.0875	3.8RP 3.45/10.4
5-C	5/95	Light reddish purple	0.251	0.4RP 5.54/8.6
6-A	100/0	Blackish purple	0.006	1.5P 0.47/1.7
6-B	33/67	Strong blue	0.085	6.0PB 3.40/9.0
6-C	5/95	Light purplish blue	0.332	5.0PB 6.26/6.8
7-A	100/0	Very dark greenish blue	0.004	4.0B 0.31/4.3
7-B	33/67	Strong bluish green	0.121	2.0BG 4.01/10.4
7-C	5/95	Brilliant bluish green	0.342	4.5BG 6.34/8.9

[a] Pigments: 1, zinc yellow; 2, chrome yellow medium; 3, molybdate orange; 4, bon red dark; 5, "Monastral" violet R (quinacridone); 6, indanthrone blue lake; 7, "Monastral" green (phthalocyanine) [7.7]

[b] Titanium white [7.7]

saturation (roughly represented by purity) when a white pigment is added to a colored pigment. Of particular interest is series 6, which shows an increase of purity when the two pigments are mixed (from point 6-A to point 6-B). After reaching a maximum purity, shown by the hairpin turn, the purity decreases on further mixture with the white pigment. Similar hairpin patterns can be obtained with other pigments [Ref. 7.20, Fig. 10]. The change of lightness on dilution with white is indicated by the tabulated values of the luminance factor Y (Table 7.3).

Plate V shows several types of mixture lines obtained from mixtures of pigments. Mixture lines I, II, and III represent sets of mixtures of two pigments, neither of which is white. The samples shown for mixture line I are identified in Table 7.4. The tabulated color names show the wide range of green hues available when the two pigments are mixed.

Curve II in Plate V, presented by Evans in [Ref. 2.1, Fig. 18.7], is the mixture line for zinc yellow J and deep cadmium red R artists' oil paints. The mixture line given by curve III is from an article by S.R. Jones on artists' pigments employed in the past [Ref. 7.21, Fig. 12]. It represents the broad gamut of greens obtained in oil by mixing pigments Prussian blue P and lead chromate (chrome yellow) K. The curved mixture line demonstrates that, although colors K and Q are complementary with respect to CIE ILL C, no mixture of the pigments on a palette will produce a neutral gray. Jones points out that the pigment mixtures represented in this range enabled 19th-century painters for the first time to represent closely the greens found in nature. Curve IV from the same article [Ref. 7.21, Fig. 7] is a hairpin

Table 7.4. Identification of yellow and blue pigment mixtures and the colors of glossy samples reproduced in Plate V

Sample	K/B Pigment weight ratio Yellow (K)[a] Blue (B)[b]	ISCC-NBS color names	Luminance factor Y	Munsel notation Hue Value/Chroma
I-A	56/44	Very dark bluish green	0.014	2.0BG 1.75/4.6
I-B	64/36	Very dark bluish green	0.018	10.0G 1.39/6.2
I-C	75/25	Very dark green	0.029	6.5G 1.90/6.9
I-D	80/20	Deep green	0.038	4.5G 2.26/7.4
I-E	89/11	Deep yellowish green	0.067	2.0G 3.03/8.3
I-F	93/7	Deep yellowish green	0.094	1.0G 3.58/8.3
I-G	98/2	Strong yellow green	0.189	7.5GY 4.90/9.4
K	100/0	Vivid yellow	0.484	6.5Y 7.35/12.5

[a] Shading yellow (chrome yellow) [7.7]

[b] Milori blue (iron blue) [7.7]

mixture line for Prussian blue P and white lead C in oil; it resembles the mixture line for samples 6-A, 6-B, 6-C, and white (point C) in Plate IV.

These illustrations show that, when only the chromaticities of the colors of two pigments are known, the chromaticities of the colors of their mixtures cannot in general be predicted satisfactorily.

7.9 Color Images in Television, Pointillism, Four-Color Printing, and Photography

The visual mixing of colors that occurs when color-television images or pointillistic paintings are viewed is referred to in Sect. 5.8. It is important to recognize that the result of visually mixing two colors is, in such cases, represented by straight mixture lines on the chromaticity diagram. Thus, in color television, in which rays of green G and red R light are emitted simultaneously, in one kind of television set (or very rapidly, in sequence, in another kind of set), from tiny neighboring G and R phosphor dots on the screen, the chromaticity of the resulting uniformly mixed color is represented by a point on the straight mixture line that connects the points for G and R on the chromaticity diagram (Fig. 7.12). The exact location of the point on the line, such as that for yellow J (Fig. 7.12), is determined by the relative amounts of G and R in the mixture. If a yellowish green Q were desired, then the amount of G would have to be greater than the amount of R.

The full range of chromaticities available by the use of the three television phosphors, B (blue), G (green), and R (red), in Fig. 7.12 is represented by the area within the three straight lines that form the triangle (Sect. 7.3).

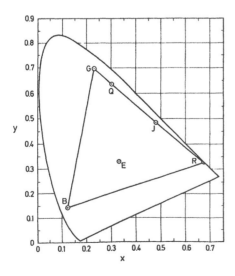

Fig. 7.12. Chromaticity gamut of colors available with three television phosphors: red (R), green (G), and blue (B) (U.S. Standard 1951) [7.22]. CIE 1931 (x, y) chromaticity diagram. Equal-energy light source E

Thus, if R and G were presented in the required relative amounts to produce Q and light from phosphor B were added, the resulting color (color mixture by averaging) would be represented by a chromaticity point within the triangle – that is, at a position on a straight line drawn between Q and B. Similarly, achromatic (white) light, such as that represented by the equal-energy source E, can be produced by a mixture of appropriate relative amounts of B, G, and R light.

The chromaticities of the colors R, G, and B are shown with those of the colors of another set of television phosphors R', G', and B' on the CIE 1976 (u', v') chromaticity diagram (Fig. 7.13). Here the two triangular ranges are shown together with the range Q of chromaticities of the present-day maximum gamut of real surface colors determined by M.R. Pointer in a survey involving over 4000 high-purity-color samples of nonfluorescent dyes and pigments [7.24]. Comparison with the surface-color range shows that one set of phosphors "overshoots", producing blue colors of purity exceeding that normally found in blue colorants; the other "overshoots" in the green blue and green domains. Comparison of the relative positions of the two triangles shows that one set of phosphors includes more colors in the green, green blue, and red domains; the other set includes more of the blue and purple domains. One set is only slightly superior in its coverage of yellows and oranges.

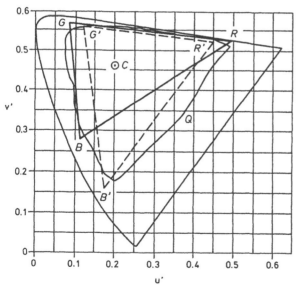

Fig. 7.13. Chromaticity gamuts of colors available with: television phosphors R, G, B (U.S. Standard 1951) (Fig. 7.12) [7.22]; television phosphors R', G', B' (European Standard: BREMA 1969) [Ref. 7.23, p. 438]; present-day colorants (nonfluorescent) Q [7.24]. CIE 1976 (u', v') chromaticity diagram. CIE ILL C. (In part from [Ref. 7.24, Fig. 12]; reproduced with the permission of John Wiley & Sons Inc., New York)

The chromaticity diagram is pertinent to pointillistic painting, not only to identify the colors of different hues obtainable by visual color mixing but also to determine which color pairs or triads can lead to drab grayish colors. For example, a straight mixture line that passes through the central region of the chromaticity diagram warns of the possibility of gray mixtures and colors of low saturation. In Sect. 7.8, comment is made on the interesting range of green colors obtainable by subtractive color mixture of Prussian blue and chrome yellow paints (curve III, Plate V). In Fig. 7.14 the curved mixture line is shown again to emphasize how it avoids the central region near point C. If, however, these two paints were applied to a canvas in a pointillistic manner, then a straight mixture line between points P and K would be followed. These colors are essentially complementary; the mixture line passes rather close to C. Therefore, unless there is a preponderance of either yellow or blue, an area painted in this manner would have a dull grayish appearance. It is true that a pointillist would probably not consider juxtaposing chrome yellow K and very dark Prussian blue P dots. Figure 7.14 shows, however, that dots of blue tint Q or of blue tints S or T (curves III and IV in Plate V) with dots of chrome yellow K would also result in mixture lines that pass through the central region that represents the chromaticities of gray or grayish colors.

Plate VI shows a mosaic of square pieces of glossy colored paper [ISCC-NBS color names: strong blue and vivid yellow (Sect. 10.1)], where each color occupies about one-half of the colored area. The CIE specification of the colors of the two original samples from which the mosaic was made are (CIE ILL C): blue (0.184, 0.207, 0.205); yellow (0.482, 0.493, 0.613). When the mosaic is viewed from a sufficient distance so that the colors are visually mixed, the whole area takes on a grayish yellow appearance. The

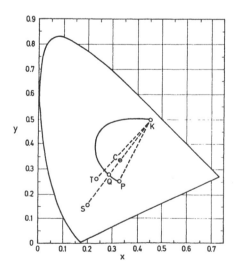

Fig. 7.14. Mixture lines for mixtures of Prussian blue (P), or blue tints (Q, S, T), and chrome yellow (K) paints. *Curved line:* subtractive color mixture; *straight dashed lines:* color mixture by averaging (pointillistic painting). See Plate V

specification of the grayish yellow determined for color mixture by averaging (Sects. 5.8, 7.3) is CIE (0.350, 0.353, 0.409) CIE ILL C. Plate VI also shows brilliant bluish green (0.221, 0.343, 0.345) and vivid purplish red (0.456, 0.259, 0.164) samples. A mosaic made of small equal squares cut from the two original samples would appear a neutral gray (medium gray) (0.310, 0.315, 0.252) when viewed at a distance.

Pointillism leads to low-purity or grayish colors if the hues of the individual dots oppose each other on the hue circle. But if two hues are close to each other on the hue circle, the colors produced through their mixture by averaging can have a purity that is greater than the purity of one of the colors. This could be demonstrated by samples 1 and 2 in Plate II. The chromaticity of any mixture (additive, or by averaging) of the two colors would be located on a straight mixture line between the points for samples 1 and 2. The point for sample 1 is closer to the point for CIE ILL C than any other point on the mixture line, hence the purity of sample 1 is the lowest. Similarly, samples 12 and 13 could be mixed visually to produce an orange of substantially unchanged purity. [But if samples 6 and 20 were visually mixed, a straight line connecting the two points (passing close to point C) would show that grayish greens or grayish violets would be produced if the proportions of green and purple violet were about the same] .

Pointillistic paintings are enjoyed not only when they are viewed at distances sufficient for complete visual mixture of color dots, but also at closer range where some of the dots are distinguishable and another phenomenon occurs (assimilation, Sect. 11.11) producing a desired vivid quality.

In printing magazines, advertising brochures, books, etc., the four-color process is frequently employed [for example, the halftone (letter press) and offset (lithography) processes] [Ref. 7.23, p. 517]. Examination of a color print (on white paper) with a magnifying glass shows that magenta, cyan, yellow, and black dots of varying size are printed. There can also be much partial covering of one dot by another, which results in the introduction of additional colors on the paper (by subtractive color mixture). The areas of the dots and the interstices between them (white) scatter light to our eyes, and we experience color mixture by averaging, much as we do when we view a pointillistic painting.

In color photographic transparencies and in color photographic prints, color results from subtractive mixture. Processed transparencies and prints consist of a sandwich of three layers, each of which contains a *dye image* (cyan, magenta, or yellow). The colors seen in transparencies result from the passage of light through the three superimposed dye images. In the case of photographic prints, the light passes through the three layers, is reflected from the white base, and is transmitted through the three layers again before it reaches our eyes.

The color gamuts available with color transparencies are quite large, much like those attained with color television [Ref. 1.39, Fig. 9.1]. In four-color printing, the gamut tends to be considerably smaller. Although a large gamut is generally very desirable, to increase color fidelity in reproduction, other factors may sometimes need to be considered, such as metamerism, chromatic adaptation, and simultaneous contrast [Refs. 1.39, p. 115; 7.23, p. 168] (Sects. 7.12, 11.4, 11.8). For example, with respect to metamerism, Hunt has shown that the introduction of spectrally more selective dyes to enlarge the color gamut of photographic film increases the likelihood that a satisfactory reproduction of a pale or dull color in one illumination will not be satisfactory in another illumination, when the original object and its photographic image are compared [Ref. 7.23, p. 168]. Human face or other skin colors may be among such troublesome colors.

7.10 Color Difference

In industry, not only is the measurement of color important but also the measurement of color differences. The reason is that in the production of large quantities of paint, fabrics, and other colored materials and objects, colors are usually required to match standards within a stated tolerance of variation; the smaller the tolerance, the more difficult the task of manufacture.

In early publications in the United States (beginning in 1940) concerning color measurements, there are numerous references to the *NBS unit* (sometimes called the *judd*) for designating small color differences. A color variation of one NBS unit represents about what is customarily tolerated in commerce [7.25,26]. One NBS unit is equivalent to a color difference that is approximately four times as great as a just-perceptible color difference. (The Munsell equivalents are given in Sect. 8.4.) At the center of the chromaticity diagram, a variation in x or y of about 0.0015 to 0.0025 corresponds to one NBS unit [Ref. 6.2a, p. 109]. Numerical values for color differences in NBS units may be calculated from the CIE tristimulus values X, Y, and Z determined for each color [Refs. 1.18, p. 317; 7.27, p. 292].

In the literature on the subject of color published during the years 1960–1975, there are discussions of a number of other proposed formulas for calculating color differences [Refs. 5.16, p. 45; 7.28]. One formula, the *ANLAB (40) formula* [Ref. 7.27, p. 31], enjoyed a rather wide acceptance, particularly in the British textile industry [7.29,30]. More recently, a simplified version of the ANLAB (40) formula was recommended by the CIE. This version, called the *CIELAB (1976) formula,* or sometimes the *CIE (1976) (L*a*b*) formula,* is recommended for universal use (textiles, paints, plastics, etc.) [7.31–34]. The formula enables the calculation of color differences from two sets of tristimulus values X, Y, and Z [Notes 7.3, 7.4]. Approximately one ANLAB (40) unit or 0.9 CIELAB (1976) unit is equivalent to one NBS unit or four times as great as a just-perceptible color difference.

The CIELAB (1976) formula is widely used for determining differences in surface colors. Another, called the *CIELUV (1976) formula,* or sometimes the *CIE (1976)* $(L^*u^*v^*)$ *formula,* is frequently used for determining color differences in applications in which additive color mixture is involved and the straight-line behavior of such mixtures on the accompanying CIE 1976 (u', v') chromaticity diagram is convenient, e.g. in lighting and television [Notes 7.3, 7.4] [3.4; 7.34,35]. Although, on average, the units of CIELAB and CIELUV color differences are approximately equal, they vary greatly for different locations and directions of difference in the (x, y) chromaticity diagram.

Color differences calculated with either of the two CIE formulas are intended to represent perceived color differences. They apply to a typical observer whose vision is adapted to daylight, viewing, for example, an object in white or neutral middle gray surroundings [2.7]. The formulas are intended for the calculation of small color differences – that is, in the range of 1–10 CIELAB (1976) units.

Some recent studies have indicated various merits and limitations of the two formulas, but at present there appears to be no clear indication of the superiority of one over the other. The CIE in a 1978 statement recommends that either formula be used "pending the development of a space and formula giving substantially better correlation with visual judgments" [7.36,37].

As an instance of color-difference measurements that are important to artists and designers, the Levison report (published in 1976) on the light-fastness of numerous artist's pigments should be noted [5.1]. Levison determined the CIE tristimulus values X, Y, and Z of the colors of paint samples (containing the pigments) before and after exposing them to light. Because there was no internationally adopted color-difference formula at the time of his investigation, he reported his results in two ways, using two formulas, one of which was the ANLAB (40) color-difference formula. Using the AN-LAB (40) color-difference results, he calculated a permanency rating for the various pigments [Note 7.5].

Most of the pigments studied by Levison exhibited changes of both chromaticity and luminance factor after exposure to light. In Fig. 6.14 the chromaticity of a yellow acrylic paint sample before (J_1) and after (J_2) several months of exposure to sunlight (in Ohio) is presented $(Y = 0.893)$. The sample (Hansa 10G, tint) is one of a number for which the chromaticity changed, but the luminance factor Y did not. The color change expressed in ANLAB (40) color units is 12.7; the corresponding permanency rating is 3.2. (The permanency-rating scale extends from zero for a marginally acceptable lightfastness for the fine arts to 10 for 100 % lightfastness.)

7.11 Colors of High Contrast

In the previous section, the measurement of small color differences was discussed briefly and an application in monitoring the color change in fading paint samples was cited. Sometimes large color differences are sought to make objects as conspicuous as possible, as in safety signs and signals, in commercial packaging, in advertising, and even in art. Let us now turn to pertinent information on colors of high contrast.

An early comprehensive study of color contrast was made by Kelly, who selected 22 colors of maximum contrast for use in color coding, for example for safety and commercial applications, from the ISCC-NBS collection of 267 centroid color chips that sample the full three-dimensional gamut of surface colors (Sect. 10.2) [7.38]. The colors, identified by their ISCC-NBS color names and centroid numbers, are to be considered in the order shown in Table 7.5. Each color contrasts maximally in hue or lightness with the one immediately preceding it in the list and contrasts significantly with earlier ones. [The first nine colors provide maximum contrast not only for persons with normal color vision but also for those with color-deficient vision or color blindness (red-green deficiency).] The numerical values of luminance factor

Table 7.5. Kelly's list of colors of maximum contrast [7.38]

Color selection number	ISCC-NBS color name	ISCC-NBS centroid number	Luminance factor Y	Munsell value V
1	White	263	0.90	9.5
2	Black	267	0.0094	0.8
3	Vivid yellow	82	0.59	8.0
4	Strong purple	218	0.14	4.3
5	Vivid orange	48	0.36	6.5
6	Very light blue	180	0.57	7.9
7	Vivid red	11	0.11	3.9
8	Grayish yellow	90	0.46	7.2
9	Medium gray	265	0.24	5.4
10	Vivid green	139	0.19	4.9
11	Strong purplish pink	247	0.40	6.8
12	Strong blue	178	0.13	4.1
13	Strong yellowish pink	26	0.43	7.0
14	Strong violet	207	0.10	3.7
15	Vivid orange yellow	66	0.48	7.3
16	Strong purplish red	255	0.15	4.4
17	Vivid greenish yellow	97	0.63	8.2
18	Strong reddish brown	40	0.070	3.1
19	Vivid yellow green	115	0.40	6.8
20	Deep yellowish brown	75	0.070	3.1
21	Vivid reddish orange	34	0.24	5.4
22	Dark olive green	126	0.036	2.2

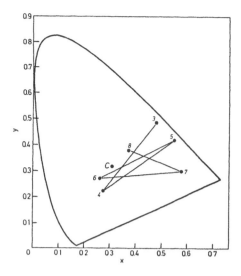

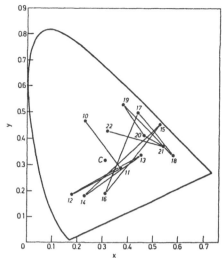

Fig. 7.15. Color samples of maximum contrast (samples 3–8, Table 7.5) [7.38]

Fig. 7.16. Color samples of maximum contrast (samples 10–22, Table 7.5) [7.38]

and Munsell Value (Sect. 8.4) given in Table 7.5 indicate the variations of lightness. Figures 7.15 and 16 show lines that join the chromaticity points of the colors, in the order listed. The closeness of a line to the point for CIE ILL C indicates the degree to which the pair of colors is an additive complementary color pair. Those pairs of colors that are not complementary, exhibit predominantly lightness contrasts.

Kelly's list is a rather specific one in that, if five contrasting colors are needed in a given application, use of the first five on his list is recommended; if 10 are needed, the first 10 are suggested, and so on. Kelly's criterion was that the farther apart two points are in color space, the more discriminable they are. This criterion has been supported recently by R.C. Carter and E.C. Carter [7.39]. They point out, however, that Kelly's list does not provide maximum discriminibility for any given number of colors. A better selection might be made for five, or for 10. In this sense, Kelly's list could be regarded as a compromise for a series of ranges, up to a range of 22 colors.

R.C. Carter and E.C. Carter investigated a color gamut provided by a laboratory color cathode-ray tube (CRT), a gamut substantially different from that considered by Kelly. With the aid of a computer, they determined a set of six colors such that the difference between any two was as large as possible. Three of the colors (red R, green G, and blue B) were those produced individually by the three phosphors. Two colors were produced by two phosphors: reddish purple rP and yellowish green yG; one color, bluish gray bGy, was produced by the three phosphors. Each color (except bGy) is located on the gamut boundary – that is, on the surface of the color solid defined by the chromaticities of the colors of the phosphors and by their luminance ranges. The minimum difference between two colors was 124 CIELUV units, which is easily discriminable. They found that a difference of about 40 CIELUV units or less is too small to be generally useful in high-contrast applications.

Some color pairs of high contrast produce an uncomfortable sensation, which artists call "vibration" [Ref. 1.17, p. 130]. Vibration occurs when the colors are complementary or near complementary, of moderate or high purity, of about equal lightness, and are situated immediately adjacent to one another. Artists employ complementary colors often to produce effects of illumination, but usually they avoid juxtaposing complementary colors of equal lightness. Many of the colors listed consecutively in Table 7.5 and indicated in Figs. 7.15 and 16 show wide variations of both luminance factor Y and purity. No vibration is expected for any of the indicated consecutive pairs.

7.12 Metamers

The subject of metamers was introduced in Sect. 5.5 with an example concerning the matching of paint. Now, having considered the CIE chromaticity diagram and the role of metamers in the CIE system (Sects. 6.2,3,6), we are in a position to discuss metamers on a more quantitative basis.

Let us begin by considering samples of two paints that contain different pigments but that match perfectly when viewed in daylight. The light scattered to our eyes from each of these samples has a different wavelength composition, because the light-absorption characteristics of the paint samples are different. Figure 7.17 shows two different hypothetical spectral reflectance curves that could apply to two such samples, P and Q.

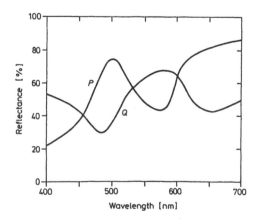

Fig. 7.17. Hypothetical spectral reflectance curves for two paint films, P and Q, that match in daylight (CIE ILL C)

The example of P and Q, although extreme, is not an unusual example of metamerism. The color stimuli (the light coming to our eyes from the two samples) are metamers; they have different spectral power distributions (Sect. 4.5) that nevertheless produce the same color under the same daylight viewing conditions. The two spectral reflectance curves responsible for the differences in the spectral power distributions show chracteristically three (sometimes more) crossover points [Refs. 6.2, p. 146; 7.40].

The color in daylight of the two samples has one CIE specification: CIE 1931 (0.343, 0.353, 0.573) CIE ILL C. The chromaticity of the color is indicated by point PQ on the chromaticity diagram in Fig. 7.18. The dashed line extending from point C (representing the chromaticity of CIE ILL C)

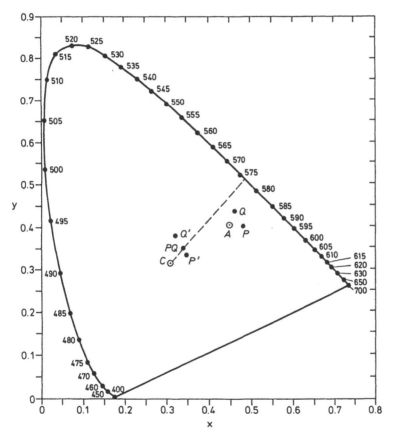

Fig. 7.18. CIE 1931 (x, y) chromaticity diagram illustrating metamerism. In daylight (C), the grayish paint films of Fig. 7.17 match. The chromaticity of their color is given by point PQ (CIE ILL C). In incandescent-lamp light (A), before adaptation, the films do not match: the chromaticities of their colors are given by points P and Q. After adaptation to A, the colors have the same appearance as the colors (chromaticities P' and Q') of materials viewed by the observer adapted to daylight (C) (but the films still do not match)

through point PQ intersects the spectrum locus at about 576 nm, the dominant wavelength, which corresponds to yellow in the spectrum (Table 4.2). In the ISCC-NBS system of color names of materials, the color is designated grayish yellow.

When samples P and Q are compared in the illumination provided by an ordinary tungsten-filament lamp A (closely approximated by CIE ILL A), they no longer match. That the colors of both of them are now different is suggested by their chromaticity points P and Q in Fig. 7.18. These two points apply to the colors perceived while the eye of the standard observer is still adapted to CIE ILL C, before it has become adapted to CIE ILL A. Several minutes later, after chromatic adaptation has occurred (the sensitivity of the eye has changed), there is then the tendency to see colors closer to how they are seen in daylight.

Points P' and Q' represent the chromaticities [x', y' in Note 7.6] of *corresponding colors,* which are colors (of, say, entirely different materials) that, when viewed by an observer adapted to one illumination (here CIE ILL C), have the same appearance (of hue and saturation) as the colors (whose chromaticities are represented by points P and Q [x, y in Note 7.6]) that are viewed by the same observer adapted to a second illumination (here CIE ILL A) [Ref. 4.4, p. 201]. After chromatic adaptation to CIE ILL A has occurred, one sample (in effect shown by Q') appears to have a slightly greenish cast in comparison with its original color (PQ), and the other (P') appears to have a slightly reddish cast. A further discussion of chromatic adaptation is presented in Sect. 11.4.

In commerce, especially in the textile industry, there is much interest in the *degree of metamerism.* If, for example, two dyed fabrics that match in daylight exhibit relatively large differences of color in lamplight, they are said to have a high degree of metamerism. There is, in the textile industry, a widespread effort to hold the degree of metamerism down to a practical limit. The CIE has recommended a *metamerism index* for quantitatively designating the degree of metamerism. The procedure requires two samples whose colors match in illumination typified by CIE ILL D_{65}. The index is simply the difference of the colors produced by the two samples when they are illuminated by a standard tungsten-filament lamp (CIE ILL A) [Refs. 4.4, p. 122; 7.41; 7.42]. Either of the CIE color difference evaluations (CIELAB, CIELUV) may be used (Sect. 7.10) [Note 7.4].

7.13 Metameric Illumination

The topic of the previous section concerns *two selective surfaces* (two colored paint films) that possess different spectral reflectance distributions (Fig. 7.17) and that match perfectly *in one illumination,* but not in other illuminations. In the case of one illumination, the two stimuli that initiate the process of color vision are metamers, because the colors perceived are identical (they match), even though the wavelength compositions of the stimuli are different. The wavelength compositions must be different because the two paint films have different light-absorption chracteristics and hence different spectral reflectance distributions (Sect. 5.2).

The topic in this section concerns *one selective surface* whose color appearance is different in *two different illuminations*. The two illuminations are chosen such that the color of a nonselective surface (for example, a white or neutral gray sheet of paper) appears exactly the same in either. As an example of a selective surface, let us consider the peel of a lemon. In one illumination, chosen to be daylight, we know that the lemon will appear yellow. In the second illumination, the lemon will appear reddish orange. The light providing one illumination and the light providing the other are metameric stimuli (this is *metameric illumination*), because, although their wavelength compositions are different, their perceived colors are identical: they are both "white" light. Colored objects viewed in one "white" illumination and then in the other "white" illumination will generally exhibit differences in their color appearance.

We can, perhaps, find a daylight fluorescent lamp whose light matches that of a phase of daylight (represented, say, by CIE ILL D_{65}), such that a sheet of white paper in either illumination appears identical in color. The stimuli are metamers. What is of practical interest is how the color appearance of colored (selective) surfaces change when viewed in one illumination and then in the other. [Such changes of color pose a real problem in illumination engineering (Sect. 7.14).] The change of color of an object viewed in daylight and in the light from the fluorescent lamp can be determined by calculation. (Because both light sources are "white" or "near-white", no correction needs to be made for chromatic adaptation.)

For purposes of illustration, let us now consider a more pronounced effect produced by two possible (but impractical) "white" metameric illuminants. One is the equal-energy source E and the other is a mixture of an additive complementary color pair (with respect to E) of spectral (monochromatic) lights: 490 nm (blue green) and 600 nm (reddish orange) (Fig. 7.1). "White" light is produced by a mixture in the proportions (radiometric) of 2.148 units of 600-nm light and 0.877 units of 490-nm light [Ref. 6.7, p. 825].

Now let us consider the appearance of a lemon illuminated by each of the two lights. With illuminant E, the lemon's color will be yellow, because E's uniform spectral power distribution is roughly an average of that for daylight (CIE ILL C). But, in the mixture of monochromatic lights, the lemon will appear reddish orange. To find the reason why, we consult the spectral reflectance curve for the lemon (Fig. 7.19). The light supplied to the surface of the lemon consists of two wavelengths, 490 nm and 600 nm, only. The spectral reflectance curve shows that 90 % of the 490-nm light is

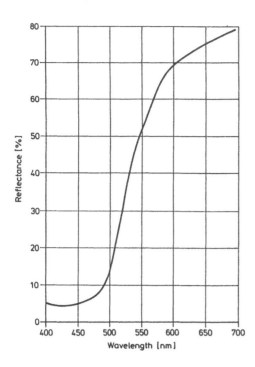

Fig. 7.19. Spectral reflectance curve for a lemon. (Courtesy of R.G. Kuehni, Mobay Chemical Corporation, Rock Hill, South Carolina)

absorbed, whereas only 31 % of the 600-nm light is absorbed. Thus, what is scattered by the lemon's pigments and sent to our eyes is mostly 600-nm light. The reddish orange perceived is less saturated than that perceived in a direct beam of 600-nm spectral light because of the addition of an achromatic component [equivalent to the additive color mixture of all the scattered 490-nm blue green light with enough of its complementary color (600-nm reddish orange light) to "neutralize" it chromatically].

A practical means of producing achromatic metameric illumination requires the use of projectors equipped with light absorption filters and a means for controlling light intensity. Slide projectors [7.43] or stage-lighting equipment [7.3] may be employed, depending upon the size of the visual display desired. Filters may be available [Refs. 3.14, pp. 71, 143; 4.4, p. 108; 7.3] that produce beams whose dominant wavelengths are acceptably close to, say, 490 and 600 nm. The dominant wavelength of the color of the lemon in the illumination produced by two such beams will not necessarily be 600 nm, and the resulting perceived saturation will be less than if monochromatic lights are used. As pointed out in Sects. 5.5 and 7.4, metameric illumination may be used to produce interesting effects in light art and stage lighting.

7.14 Color Rendering

In the previous section, attention was drawn to the marked changes of color appearance that are possible when colored objects are viewed in different achromatic ("white") illuminations that are of the same metamer set. The purpose was to illustrate unexpected effects that can be produced. On the other hand, in everyday life it is far more important to know which artificial illuminations can be employed without producing objectionable changes of color appearance relative to a perceived color in daylight (or, sometimes, in incandescent-lamp illumination). Artists confront this problem when they work by north-sky light by day and by artificial illumination by night. Similarly, decisions have to be made about the best illumination for displays of works of art in museums. The color quality of illumination is important in displays of food, for the performance of certain tasks in industry, in hospital operating rooms, and in many other professional, industrial, and agricultural situations.

Following the introduction of the modern fluorescent lamp in 1939, there has been a rapid development of efficient light sources. Along with this development, much attention has been given to the quality of illumination. This has been possible because, in the design of fluorescent lamps, the color temperature (Sect. 7.15) and the spectral power distribution of the radiation (Fig. 4.2) can be varied over wide ranges [Refs. 7.23, p. 160; 7.44, p. 219; 7.45]. As a consequence, a variety of fluorescent-lamp types are commercially available. Some have been designed for economical operation, others for quality of illumination (specifically concerning *color rendering*).

The term "color rendering" applies to the "effect of an illuminant on the color appearance of objects in conscious or subconscious comparison with their color appearance under the reference illuminant" [7.46,47]. A *color-rendering index* has been devised to serve as a "measure of the degree to which the perceived colors of objects illuminated by the source conform to those of the same objects illuminated by a reference illuminant for specified conditions".

The reference illuminant for daylight lighting is a CIE illuminant corresponding to a phase of daylight [Ref. 4.4, p. 98] chosen such that the difference between the chromaticities of the colors of the reference illuminant and the test illuminant is a minimum. Color differences for each of eight standard color samples in the two illuminations are calculated. From an average of these color differences, a *general color-rendering index* R_a is determined [Notes 7.4 and 7.7] [7.48]. (The eight color samples have different hues and moderate Munsell Value and Chroma; they are designated approximately as follows: 7.5 R 6/4, 5 Y 6/4, 5 GY 6/8, 2.5 G 6/6, 10 BG 6/4, 5 PB 6/8, 2.5 P 6/8, 10 P 6/8 [7.47].)

A general color-rendering index rating of 100 implies zero color differences for all eight samples and indicates complete color rendering. For demanding visual tasks such as color matching, the illumination provided should have an index of 95 or more, but a minimum index of 91 is tolerated [Ref. 7.44, p. 222]. The CIE warns that lamps of the same R_a are not necessarily interchangeable. This problem would occur, for example, in cases for which the color differences found using one lamp and the reference were equal but opposite to those found using a second lamp and the reference.

The general color-rendering index R_a applies to the illumination of a collection of diversely colored objects. When the object or objects being illuminated are more restricted in color range, a *special color-rendering index* R may be devised that employs only the appropriate samples out of the collection of eight. In addition to the eight color samples, the CIE recommends six others for possible use in a special color-rendering index. In Munsell notation they are (approximately): 4.5 R 4/13, 5 Y 8/10, 4.5 G 5/8, 3 PB 3/11, 5 YR 8/4 (caucasian complexion), 5 GY 4/4 (leaf green) [7.47].

Typical color-rendering indexes of various lamps have been published in books by P.J. Bouma [Ref. 7.44, p. 233] and by Hunt [Ref. 7.23, p. 162]. Bouma recorded a low general color-rendering index ($R_a = 67$) for a standard cool-white fluorscent lamp. The special color-rendering index rating for the purple sample was low ($R = 40$), indicating a poor match, and a poor match was found also for the red sample ($R = 58$). But the rating for green yellow was excellent ($R = 99$). This shows that a low index may be due to very erratic (rather than uniformly poor) behavior. Another fluorescent lamp of the same general index R_a could show entirely different variations.

On the other hand, a "super de luxe" cool-white fluorescent lamp had a very high general index ($R_a = 97$). The special index ratings for each of the eight samples varied over a narrow range ($R = 94$ to 99). This lamp is of the type required for faithful color rendering under demanding circumstances. For general use, a general color-rendering index rating of $R_a = 80$ to 85 is deemed acceptable. It is found that lamps of high power efficiency tend to have lower ratings. The most efficient commercial lamps (high-pressure sodium lamps) have a very low general color-rendering index (say, $R_a = 21$) [Ref. 7.23, p. 162].

7.15 Color Temperature

Very frequently in the literature on color, the term *color temperature* is encountered (or its equivalent, *blackbody temperature*), particularly in the specifications of the colors of light emitted by lamps. To illustrate the concept briefly, let us consider a blackbody and its color (of incandescence) at various temperatures.

A blackbody is a theoretical object, but it can be approximated well in a physics laboratory. The laboratory object could be a block of a refractory material (ceramic) having a closed cavity within it. When the block is maintained at some uniform temperature, say 1000°C, in an electric furnace, for example, the color of the interior wall of the cavity is the color of a theoretical blackbody at 1000°C, the blackbody temperature. If a small hole is drilled into the block, a hole small enough not to alter significantly the uniform radiation within the cavity, then a portion of the interior wall of the hot cavity can be observed and its color measured at the blackbody temperature. The color does not depend on the material used for the block.

When the temperature of the cavity wall is about 500°C, it is a dull red; at 750°C it is a reddish orange. It is a brillant orange at the melting point of iron (1535°C). If the experiment is continued to higher temperatures, the color becomes white at 3000°C and would be bluish white if the temperature could be raised to 10 000°C. The points that represent the chromaticities of the cavity wall as it is heated fall rather closely along the dashed curved line shown in Fig. 7.20, starting at the lower-right corner of the CIE chromaticity diagram. A portion of the diagram containing the curve is reproduced on a larger scale in Fig. 7.21. The curve is called the *color-temperature curve*, or sometimes the *blackbody locus* or the *planckian locus*.

At this point, it is appropriate to introduce the Kelvin temperature scale, which is commonly employed in the physical sciences. It is used in all designations of color temperature. The Kelvin scale is simple; its zero represents the lowest possible temperature. The temperature divisions on the Kelvin scale are identical with those on the Centigrade (Celsius) scale. Thus a change of 5°K and a change of 5°C are exactly the same. On the Centigrade scale, the lowest possible temperature is −273.16°C. Therefore, to convert a temperature in degrees Centigrade to degrees Kelvin, we simply add 273.16 to it. In a recent international agreement on scientific terminology, the term *degrees Kelvin* (°K) was replaced by the term *kelvin* (K). From the above we see that the temperature of 3000°C is converted to 3273 K, and a temperature given roughly as 4000°C is adequately represented by 4300 K.

The color-temperature curve shown in Fig. 7.21 could have been established readily and precisely by laboratory experiment only in the lower

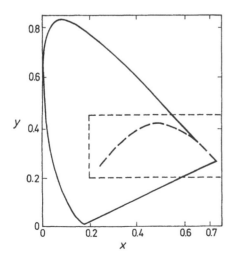

Fig. 7.20. Color-temperature curve (*curved dashed line*). The area enclosed by the *straight dashed lines* is enlarged in Fig. 7.21

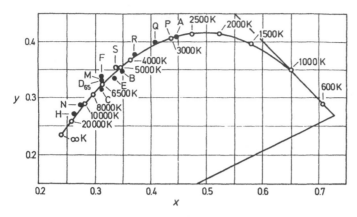

Fig. 7.21. Color-temperature curve [Ref. 1.18, Table 2.13]. Enlargement of a portion of Fig. 7.20 showing the full range of the color-temperature curve in kelvins [K] up to infinite (∞) color temperature. Chromaticity points are shown for: CIE ILL A,B,C and D_{65} [Ref. 1.18, p. 166]; equal-energy light (E); light from 40 W standard warm white (P), white (Q), standard cool white (R), and daylight (F) fluorescent lamps [Ref. 3.14, p. 47]; light from an overcast sky (M) [4.17]; north-sky light falling on a 45° plane (N) (curve *II*, Fig. 4.3) [4.17]; direct sunlight (S) (curve *I*, Fig. 4.3) [4.17]; and light from a clear blue sky (H) [Ref. 3.14, p. 47]

temperature range. In fact, the curve was established by a precise mathematical means based upon a theoretical equation that relates the spectral power distribution of a blackbody's thermal radiation to its temperature [Ref. 4.4, p. 27]. [The relative spectral power distribution curve for CIE ILL A (Fig. 4.1) was determined with the use of the equation; it is the relative spectral power distribution curve for a blackbody at 2856 K.] The color-temperature curve extends through the entire practical range up to the theoretical limit of infinite temperature (∞K).

It should be emphasized that the color-temperature curve precisely relates temperature to color only for a blackbody. The curve is used extensively by color scientists and illumination engineers because the chromaticities of the colors of the light produced by most of the lamps used today fall on or near it. The approximate chromaticity (color quality) of lamplight can be specified simply by stating its color temperature (or its correlated color temperature, discussed below). Color temperature is easier to use and more suggestive of color quality than is CIE (x, y) chromaticity notation. Compare, for example, a report that the color temperature of lamplight is 6500 K and that its chromaticity is (0.313, 0.329). With a little experience, the significance of what 6500 K represents in color appearance is more readily grasped.

The open points plotted along the color-temperature curve in Fig. 7.21 show the chromaticities of the colors of a blackbody at the indicated temperatures. What is important here, however, is that the open points provide a color-temperature scale: 1000 K, 1500 K, 2000 K, ..., 20 000 K. The black dots are the chromaticity points of the colors of light from various white and near-white light sources. The position of a black dot *relative to the scale* enables us to estimate the color temperature of light from a lamp. In this way, the color temperature of light from a fluorescent lamp, of CIE ILL B, and of north-sky light can be found. Color temperatures bear no relation to the actual temperatures of such sources.

Most black points shown do not fall precisely on the curve, however. In these cases, for more precision, a *correlated color temperature* is employed. Let us consider any one black point that is not on the curve. The correlated color temperature corresponding to that point is the color temperature of a blackbody whose chromaticity appears most-nearly similar to it. The most-nearly similar blackbody chromaticity could be determined on a truly uniform chromaticity diagram, but, lacking such, the task is conveniently accomplished using a published chart [Ref. 1.18, Fig. 2.25].

Direct sunlight (B, S), equal-energy light (E), daylight (C, D_{65}), light from an overcast sky (M), and north-sky light (N) indicated in Fig. 7.21 are commonly considered "white". Evans described the white region of the chromaticity diagram as a sausage-shaped area that includes the color-temperature curve from about 4000 to 10 000 K [Ref. 2.5, p. 51]. Although the length of the "sausage" is indicated, the only reference to its width is "the distance to either side of the color-temperature curve being shorter

Table 7.6. Chromaticity and color temperature. CIE 1931 (x, y) and 1964 (x_{10}, y_{10}) chromaticities for several CIE illuminants, equal-energy light E, and examples of daylight. (See Table 12.4 for CIE 1976 (u', v') and (u'_{10}, v'_{10}) chromaticities for CIE illuminants and equal-energy light E)

	CIE 1931 chromaticity coordinates		CIE 1964 chromaticity coordinates		Color temperature [K]
	x	y	x_{10}	y_{10}	
CIE ILL A (typical of incandescent lamplight) [Ref. 1.18, p. 166]	0.4476	0.4074	0.4512	0.4059	2856
CIE ILL B (typical of direct sunlight) [Ref. 1.18, p. 166]	0.3484	0.3516	0.3498	0.3527	4874[a]
CIE ILL C (typical of average daylight) [Ref. 1.18, p. 166]	0.3101	0.3162	0.3104	0.3191	6774[a]
CIE ILL D_{65} (typical of average daylight) [Ref. 1.18, p. 166]	0.3127	0.3290	0.3138	0.3310	6504[a]
Direct sunlight [4.17]	0.3362	0.3502			5335[a]
Light from overcast sky [4.17]	0.3134	0.3275			6500[a]
Light from north sky on a 45° plane [4.17]	0.2773	0.2934			10,000[a]
Light from equal-energy source E	0.3333	0.3333	0.3333	0.3333	5400[a]

[a] Correlated color temperature

103

across than along it." The "sausage" (dashed oval) in Fig. 7.1 is a guess for the shape of the region. At color temperatures that exceed 10 000 K, the corresponding colors are distinctly bluish whites, of increasing purity.

In Table 7.6 the chromaticities CIE 1931 (x, y) (based on a 2° angle of vision) and CIE 1964 (x_{10}, y_{10}) (based on a 10° angle of vision) and the color temperature (or correlated color temperature) are given for the equal-energy source E; CIE ILL A, B, C, and D_{65} (Figs. 4.4,5); light from several fluorescent lamps; and light from the sky under three conditions (for two, see curves I and II, Fig. 4.3). The chromaticity point for the light from a standard clear 60-W tungsten-filament light bulb is approximately coincident with point A (CIE ILL A) (Fig. 7.21).

As mentioned above, except for a blackbody, color temperature does not reveal the temperature of the source. For example, the light from a tungsten-filament lamp operating at 2500 K has a color temperature that is somewhat higher. The color temperature of sunlight observed at the surface of the earth is somewhat lower than the actual temperature of the surface of the sun (about 6000 K). The color temperature of the blue sky (10 000–15 000 K) is a function of a wavelength composition resulting from scattering of sunlight by molecules in the atmosphere (Sect. 2.2); air temperature is not a deciding factor. In a fluorescent lamp, light is emitted by phosphors because of bombardment by a beam of electrons; the light is not caused by high temperatures. The color temperatures of light from fluorescent lamps are very much higher than their temperatures of operation.

8. Color Systems

8.1 CIE Color Space, CIE(x, y, Y)

It is often said that color is three-dimensional. (This is true at least for psychophysical color and isolated psychological color.) But what is meant by the *three dimensions of color*? Commonly, we think of the word "dimensions" in terms of height, width, and depth, all stated in feet, meters, or other units of length. All objects have volume and occupy space; they are three-dimensional.

The dimensions of color are the quantities that specify it. The dimensions of isolated psychological color are hue, saturation, and brightness. Those of psychophysical color are the CIE tristimulus values X, Y, and Z or three independent quantities derived from them, such as x, y, and Y, or the set λ_D, p_e, and Y, or the sets mentioned in Sect. 8.2.

The chromaticity diagram, which is concerned with only x and y, is two-dimensional; it can be printed on a flat piece of graph paper. However, in order to represent, simultaneously, the three dimensions x, y, and Y graphically, the points must be plotted in space – that is, CIE(x, y, Y) color space. Such color space can be visualized as consisting of a series of horizontal chromaticity diagrams arranged one precisely above another, like a series of floors in a high-rise building. Each chromaticity diagram in the series would accommodate points that represent colors of a single luminance Y. Thus, at one level, luminance Y would be 50, for example; at the next level above, 60; at the next, 70; etc. (Fig. 8.1). We could, of course, imagine more closely spaced levels, such as $Y = 50, 51, 52, 53$, etc., with a continuous variation of Y between the levels, permitting a color of $Y = 52.6$ to be located precisely on an intermediate level situated six-tenths (0.6) of the distance between levels 52 and 53 above level 52.

It should be noted that, whereas only the chromaticity (x, y) of a color is represented on a chromaticity diagram (two-dimensional), in color space color (x, y, Y) itself is represented. (In the case of colored objects, the illuminant must also be known to complete the specification of a color.)

If a beam of light A of one color (for which $Y = 50$) is combined with a beam B of another color (for which $Y = 60$), the resulting additive mixture M would have a luminance $Y = 110$. In CIE(x, y, Y) color space, the chromaticity point for the color of beam A would be at the level $Y =$

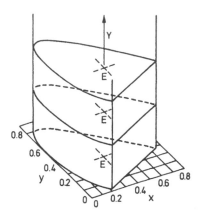

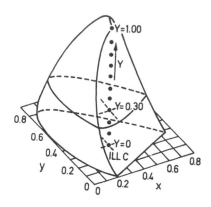

Fig. 8.1   **Fig. 8.2**

Fig. 8.1. CIE 1931 (x, y, Y) color space for light emitted by luminous objects

Fig. 8.2. CIE 1931 (x, y, Y) color space (defined by the MacAdam limits) for light scattered by or transmitted through nonfluorescent, non-self-luminous objects illuminated by light from a source typified by CIE ILL C. (Based on [Ref. 3.14, Fig. 3.22]; reproduced with the permission of John Wiley & Sons Inc., New York)

50, that of B would be at the level $Y = 60$, and that of M would be at $Y = 110$. The location of the chromaticity point for M at $Y = 110$ would be determined by the method explained in Sect. 7.3.

For the colors of *light*, CIE(x, y, Y) color space extends upward to a luminance level beyond which the light is dazzling and cannot be tolerated. Figure 8.1 shows the color space for *light* emitted by luminous objects, extending from $Y = 0$ to the luminance corresponding to the tolerance limit.

Colors of nonluminous *objects* can be represented by points in color space where Y is the luminance factor, but, because of light absorption and the wavelength composition of the illuminant employed, the points that represent nonfluorescent colors are confined to a more limited color space. This color space is defined by the MacAdam limits (Sect. 7.5), as shown in Fig. 8.2 for illumination typified by CIE ILL C.

In the case of the colors of objects, we must imagine the color space to have a structure more like a pyramid than a conventional high-rise building. The floor plan at each level (corresponding to the luminance factor Y) would have the shape outlined by the MacAdam limit. The floor plans at $Y = 0, 0.10, 0.20, \ldots, 1.00$ are shown in Fig. 7.6. The complete pyramid has the form shown in Fig. 8.2 in which one contour (or level) is indicated at $Y = 0.30$. Plate II shows samples of colors that are located at about the level of $Y = 0.30$ in CIE(x, y, Y) color space. The fact that the bottom plane $Y = 0$ of the color solid (Fig. 8.2) represents one color only, black, has been discussed in Sect. 7.5. In some of the color spaces described below, black is represented by a bottommost point (for example, CIELUV color space, Munsell color space, and DIN-6164 color space) [Note 7.1]. Another CIE 1931 (x, y, Y) color space for which CIE ILL A is the illuminant has also been presented by MacAdam [Refs. 3.14, Fig. 3.20; 4.4, Fig. 7.24; 4.18, Fig. 87; 7.4].

8.2 CIELUV and CIELAB Color Spaces

The CIELUV and CIELAB color spaces are recent developments intended to approximate uniform color space. The shortest distance between two points representing colors in either space is a measure of the difference of the colors (Sect. 7.10). This distance or color difference can be calculated with the use of a formula [Note 7.4].

Interest in finding a sufficiently uniform color space for the determination of color differences has intensified in recent years. The degree of uniformity of CIELUV and CIELAB color spaces is adequate for many purposes but insufficient for others, so the search continues. Let us begin with a few brief remarks concerning the historical background of these two color spaces before surveying their properties.

The CIE 1960 (u, v) chromaticity diagram (Sect. 6.5), which provides approximately uniform spacing in a plane, presented a starting point for the development of a color space called CIE 1964 $(U^*V^*W^*)$ uniform color space [Ref. 1.18, p. 324]. The formula for calculating color differences in this space, provisionally recommended by the CIE, was used widely for over a decade. Then a modification of CIE 1964 $(U^*V^*W^*)$ color space was adopted. The modified color space is called *CIELUV 1976 color space* and the formula that corresponds to it is the *CIELUV 1976 color-difference formula* [sometimes called the *CIE 1976 $(L^*u^*v^*)$ formula*] (Sect. 7.10) [Note 7.4]. The CIE 1976 (u', v') chromaticity diagram, associated with CIELUV color space, was adopted at the same time (Sect. 6.5).

To understand why a second color space, *CIELAB 1976 color space*, was also adopted, we need to note that the color-difference formula (Sect. 7.10 and Note 7.4) that corresponds to it is a simplified version of tha AN-LAB(40) color-difference equation that had been adopted by several organizations [7.33] and had been in extensive use, particularly in Great Britain. The new version is called the *CIELAB 1976 color-difference formula* (or, sometimes, the *CIE 1976 $(L^*a^*b^*)$ formula*).

The basic structures of the CIELUV and CIELAB color spaces are similar. In both, there is a vertical *metric lightness L^** (also called the *CIE (1976) lightness function*) axis passing centrally through evenly spaced horizontal planes that are subdivided into square grids containing coordinates (u^*, v^*) [CIELUV] or coordinates (a^*, b^*) [CIELAB]. Figures 8.3 and 4 show the vertical axis L^* that passes (from black to white) through the

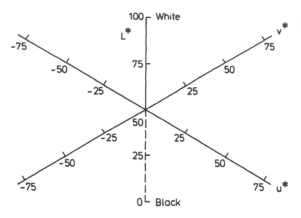

Fig. 8.3. CIELUV 1976 color space. One horizontal plane, which contains the u^* and v^* axes, is shown at a metric lightness $L^* = 50$. The vertical L^* axis cuts the plane at $u^* = 0$ and $v^* = 0$

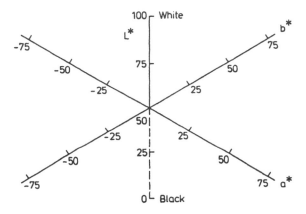

Fig. 8.4. CIELAB 1976 color space. One horizontal plane, which contains the a^* and b^* axes, is shown at metric lightness $L^* = 50$. The vertical L^* axis cuts the plane at $a^* = 0$ and $b^* = 0$

horizontal plane, (u^*, v^*) or (a^*, b^*), at $L^* = 50$, for example. Simple mathematical equations are available that relate the following [Notes 7.3 and 7.4]: (1) metric lightness L^* to luminance factor Y; (2) parameters u^* and v^* to u' and v' (which, in turn, are determined from the tristimulus values X, Y, and Z [Note 6.2]); (3) parameters a^* and b^* to the tristimulus values. [The use of positive and negative axes in these color spaces was adopted for mathematical convenience (Sect. 9.2).]

Figure 8.5 shows two points that represent colors in CIELUV color space. Point P_1 is located at $u_1^* = 12$ and $v_1^* = 26$ at level $L_1^* = 50$. A nearby point P_2 is shown at some other level of metric lightness L_2^* at a location given by different values of u_2^* and v_2^* on the L_2^* plane. The CIELUV difference between the two colors is the distance between the two points in space (CIELUV color space). Precisely the same description applies to nearby points in CIELAB color space, using coordinates (a^*, b^*) in place of (u^*, v^*).

The metric-lightness axis L^* represents hueless colors (e.g. with illuminant CIE ILL D_{65}) [3.3]. It ranges upwards from black, through the neutral

108

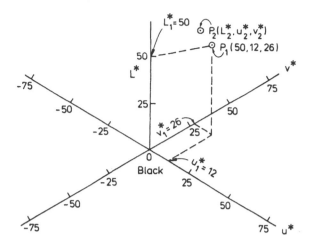

Fig. 8.5. Two colors are represented by point $P_1(50, 12, 26)$ and nearby point $P_2(L_2^*, u_2^*, v_2^*)$ in CIELUV 1976 color space. One horizontal plane, which contains the u^* and v^* axes, is shown at metric lightness $L^* = 0$. For CIELAB 1976 color space, the notations u^* and v^* are replaced by a^* and b^*, respectively

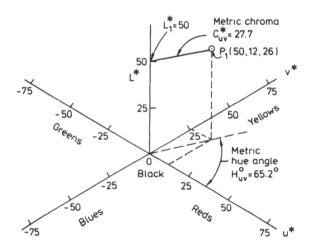

Fig. 8.6. Metric lightness L_1^*, metric chroma C_{uv}^*, and metric hue angle H_{uv}° are indicated for color P_1 in CIELUV 1976 color space. For CIELAB 1976 color space, the notations u^*, v^*, C_{uv}^*, and H_{uv}° are replaced by a^*, b^*, C_{ab}^*, and H_{ab}°, respectively

grays to white. Points in color space apart from those on the metric-lightness axis represent chromatic colors. The correlates of perceived lightness, saturation, and hue are, in both color spaces, *metric lightness L^**, *metric chroma C^**, and *metric hue angle H°*, respectively [3.3; 7.31,32]. These three variables are indicated diagrammatically for color P_1 in CIELUV 1976 color space in Fig. 8.6. The numerical value of metric chroma C_{uv} for color P_1 is the radial distance from the metric-lightness axis to point P_1 ($C_{uv}^* = 27.7$).

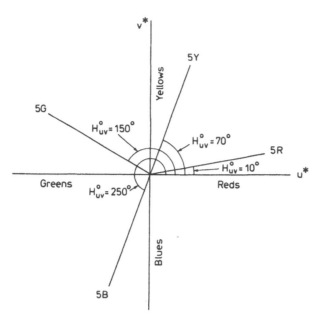

Fig. 8.7. A horizontal plane in CIELUV 1976 color space showing four examples of metric hue angle H°_{uv} (labeled by circular arcs: 10°, 70°, 150°, and 250°). These values correspond approximately to Munsell Hues 5R, 5Y, 5G, and 5PB, respectively [7.24]

The metric hue angle H° for color P_1 is the angle [degrees] measured between the metric-chroma radius and the positive (+) part of the u^* axis (or a^* axis in CIELAB color space) (Figs. 8.6,7). Metric hue angle H° is expressed on a scale from 0° to 360°, measured counterclockwise. Thus, for P_1, $H^\circ_{uv} = 65.2°$ (Fig. 8.6). Four examples of colors that have different metric hues are indicated in Fig. 8.7 in quadrants I, II, and III in a horizontal plane in CIELUV color space. In CIELUV (or CIELAB) color space, the hue regions of a horizontal plane (u^*, v^*) [or (a^*, b^*)] are suggested roughly by the directions of the axes as follows: $+u^*$ (or $+a^*$), reds; $+v^*$ (or $+b^*$), yellows; $-u^*$ (or $-a^*$), greens; $-v^*$ (or $-b^*$), blues (Figs. 8.6,7). For any given color, the numerical values of metric chroma and metric hue in CIELUV color space are generally different from those of their counterparts in CIELAB color space. For this reason, mention of the metric chroma and metric hue of a color must be accompanied by an identification of the corresponding color space (for example, by the use of subscripts: C^*_{ab}, C^*_{uv}; H°_{ab}, H°_{uv}). In both color spaces, L^* is the same quantity.

A sketch of a three-dimensional model of CIELUV 1976 color space is shown in Fig. 8.8. The top point $(L^* = 100)$ represents white and the bottom point $(L^* = 0)$ represents black. A model of CIELAB 1976 color space, defined by the CIE for color-difference measurements [Note 7.4], does not extend below $Y = 0.01 (L^* = 9)$, hence the corresponding very dark

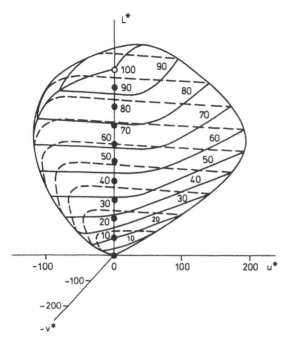

Fig. 8.8. CIELUV 1976 color space (defined by the MacAdam limits) for light scattered by or transmitted through nonfluorescent, non-self-luminous objects illuminated by light from a source typified by CIE ILL D_{65} (10° angle of vision). Metric lightness is indicated (0, 10, 20, ..., 100) along the L^* axis. Lightness planes are shown for $L^* = 10, 20, ..., 90$. (Modification of [Ref. 1.18, Fig. 2.86]; reproduced with the permission of John Wiley & Sons Inc., New York)

colors and black are excluded. (A means of extending CIELAB space from $Y = 0.01$ to $Y = 0$ has been proposed by Pauli [8.1].)

The CIELUV and CIELAB color spaces are employed as approximations to uniform color space [8.2]. Perfectly uniform color space may not be attainable in a three-dimensional euclidean model, but color experts continue to work [3.4, 8.3] towards "the development of a space and formula giving substantially better correlation with visual judgments" (Sect. 7.10).

The CIELUV and CIELAB color spaces have been developed primarily for routine color-difference measurements, but they are applied extensively in color research as well. They could be of real interest in art and design because they offer a means of specifying colors in approximately equally spaced cubical arrangements. To visualize this, imagine color space as a tightly packed arrangement of identical cubes with corners touching. Approximately equally spaced colors would be defined by the positions of the corners in color space. Each color would be approximately equally different from its six nearest neighbors. The merit of the cubical arrangement is its simplicity. Chapter 9 is devoted to a discussion of a more complex arrangement that permits each color to have 12 equally different nearest neighbors. For this system an ample collection of precise color samples has been prepared. Other important color systems that exhibit color samples in different ordering schemes are described in the following sections.

8.3 Color-Sample Systems

Although the $CIE(x, y, Y)$ system for color specification is internationally accepted and is in very active use, a number of systems that consist of *color samples* are employed in applications for which less precision is required but tangible samples are demanded. Some of these are used only in specific industries or trades (for example, textile, building, plastics, and interior decorating); in certain cases, however, the samples are identified in terms of $CIE(x, y, Y)$ specifications, which makes them generally useful. Table 8.1 lists a selection of color systems that are, or could be, of interest to artists and designers. Most of the systems provide printed, dyed, or painted samples; two provide light filters. The table cites references that give further information about the systems. The systems most familiar to artists and designers are the Munsell and Ostwald color systems, which are discussed in the next two sections; other systems of comparable importance are described in following sections. The OSA Uniform Color Scale samples, which are not intended for use in color specification, but which have other special utility in art and design, are considered in some detail separately in Chap. 9. Extensive lists of color-sample systems, old and new, have been prepared by Birren [8.5,38].

Table 8.1. Color-sample systems and color atlases [1,2,3]

1. *Chroma Cosmos 5000.* Tokyo: Japan Color Research Institute, 1978. 5000 glossy color chips. For use in design. Munsell notation; $CIE(xyY)$ notation. Text in Japanese and English. [8.4–6] (Sect. 8.8)
2. *Chromaton 707.* Tokyo: Japan Color Research Institute, 1982. 707 glossy color chips and glossy color cards. For use in design. Munsell notation. Text in Japanese and English. [8.7,8] (Sect. 8.8)
3. *Color Atlas.* E.A. Hickethier. New York: Van Nostrand Reinhold, 1974. 1000 printed samples. For color identification and notation. Conversion to $CIE(xyY)$.
4. *Color Harmony Manual.* E. Jacobson, W.C. Granville, C.E. Foss. Chicago: Container Corporation of America, 1942, 1946, 1948, 1958. 943 paint samples on cellulose acetate film, matt and glossy sides (3rd ed.). For use in design. See [8.9,10] for conversion to $CIE(xyY)$, matt samples. [Refs. 1.17; 1.18, p. 250; 1.20, p. 165; 2.5, p. 167; 6.2a, p. 182; 8.10–12] (Sect. 8.5) (publication discontinued)
5. *Color, Origin, Systems, Uses.* H. Küppers. New York: Van Nostrand Reinhold, 1973. 1400 printed samples. For the printing industry
6. *The Dictionary of Color.* A. Maerz, M.R. Paul. New York: McGraw-Hill, 1930, 1950. 7056 screen-plate printed samples on semi-glossy paper. For general use. See [8.9] for conversion (1st ed.) to $CIE(xyY)$. [Refs. 1.18, p. 252; 1.20, p. 170; 4.18, p. 337; 7.2, p. 11]
7. *DIN-Farbenkarte* (Color Chart). German Federal Republic Standard DIN-6164. Berlin: Beuth-Verlag, 1962: 590 matt paint chips; 1984: 1004 glossy paint chips. For general use. Conversion to $CIE(xyY)$ for matt and glossy chips and to Munsell notation for matt chips. [Refs. 1.18, p. 266; 3.14, p. 478; 8.13] (Sect. 8.6)
8. *Horticultural Colour Chart.* London: The British Colour Council and the Royal Horticultural Society, 1938, 1940, 1942, 1966. Copyright, Robert F. Wilson. 800 printed color samples. See [8.14] for conversion to Munsell notation. [Ref. 7.2, p. 11]

Table 8.1. Color-sample systems and color atlases (continued) [1,2,3]

9. *ICI Colour Atlas.* London: Dyestuffs Division, Imperial Chemical Industries, 1972. 1379 imprinted color swatches and 19 gray filters (27 580 color possibilities). For use in the textile industry. [Refs. 7.12, p. 13; 8.15,16]

10. *ISCC-NBS Centroid Color Charts.* Washington, D.C.: Office of Standard Reference Materials, National Bureau of Standards, 1965. 267 glossy color chips. For general use. [7.2] (Sect. 10.2)

11. *Lovibond Color.* Salisbury, England: The Tintometer Ltd. Colored glass filters. 1900 color combinations. For use in colorimetry. See [8.17] for conversion to CIE (xyY). [Ref. 1.18, p. 200]

12. *Methuen Handbook of Colour.* A. Kornerup. J.H. Wanscher. London: Methuen, 1963, 1967, 1978. 1266 printed samples (halftone). For general use. Conversion to Munsell notation. [8.18] (See *Reinhold Color Atlas*)

13. *Munsell Book of Color.* Baltimore: Munsell Color, 1929, 1942, 1973, 1976. 1325 matt color chips; 1600 glossy color chips. For general use. See [8.19–21] and Sect. 12.2 for conversion to CIE(xyY). [Refs. 1.18, p. 258; 1.20, p. 167; 2.5, p. 156; 3.14, p. 476; 4.18, p. 334; 6.2, p. 172; 8.11] (Sect. 8.4)

14. *Natural Colour System Atlas (SIS Colour Atlas NCS).* Swedish Standard No. 01 91 02. Stockholm. Scandinavian Colour Institute, 1979, 1412 matt painted chips. For architecture and design. Conversion to CIE(xyY). Text in Swedish, English, French, German, Spanish, Russian. [Refs. 1.18, p. 269; 8.22–27] (Sect. 8.7)

15. *NU-Hue Custom Color System.* Chicago: Martin-Senour Co., 1946. 1000 painted cards. For use in the paint industry. Conversion to CIE(xyY) and Munsell notation. [Refs. 1.18, p. 247; 1.20, p. 164; 8.28]

16. *OSA Uniform Color Scales.* Washington, D.C.: Optical Society of America, 1977. 558 glossy sample cards (acrylic-base paint). For use in scientific research and in art and design. See [8.29] for conversion to CIE$(x_{10}y_{10}Y_{10})$ CIE ILL D_{65} and [8.30] for conversion to Munsell notation. [Refs. 1.18, p. 270; 8.31–34] (Chap. 9)

17. *Ostwald Color System.* (See *Color Harmony Manual*)

18. *Plochere Color System.* Los Angeles: G. & G. Plochere, 1948, 1954, 1965. Colored cards: 1248 colors and 208 grayed tones. For use in interior decorating. See [8.35] for conversion to Munsell notation. [Ref. 7.2, p. 12]

19. *Reinhold Color Atlas.* A. Kornerup. J.H. Wanscher. New York: Reinhold, 1961. 1440 color samples. (See *Methuen Handbook of Color*)

20. *SCOT-Munsell System.* Newburgh, NY: Macbeth Div., Kollmorgen Corp., 1984. 2034 color samples of polyester satin-backed crepe in the SCOT book and a set of 2034 swatches. Colors identified by Munsell notation. For use in the textile industry [8.36]

21. *Standard Color Reference of America.* 10th ed. New York: Color Association of the United States, 1981. 192 dyed silk swatches (matt and shiny sides). For use in the textile industry. See [8.37] for conversion to CIE(xyY) and Munsell notation. (Formerly, *Standard Color Card of America*) [Ref. 7.2, p. 13]

22. *Villalobos Color Atlas* (Atlas de los Colores). C. Villalobos-Dominguez, J. Villalobos. Buenos Aires: Libreria El Ateneo Editorial, 1947. 7279 glossy printed chips (halftone). For color identification. [Ref. 1.20, p. 171] (publication discontinued)

23. *Wratten Light Filters.* Rochester, NY: Eastman Kodak Co., 1975. Gelatin filters. For use in photography and optical sciences. [Refs. 3.14, p. 143; 4.4, p. 107]

---

[1] All of the color-sample systems and color atlases in this table are in the Faber Birren Collection of Books on Color at the Art and Architecture Library of Yale University (Sect. 12.1), except items 10, 11, 15, 22, and 24 (1982).

[2] Aside from the various published color atlases and color-sample systems, sets of printed papers (swatches and sheets) of carefully graded colors are produced commercially for various uses. For example, a major producer in the United States is Pantone, Inc. (Moonachie, NJ).

[3] Some addresses of sources are given in Sect. 12.1.

The color systems discussed in the following sections contain rather large collections of accurately produced painted color chips. These collections (assembled in the form of color atlases) are rather expensive, and publication of one (*The Color Harmony Manual*) has been discontinued. Books that show printed samples are comparatively inexpensive and more widely accessible. The collections of painted samples are held mostly by industrial organizations, institutes of science and technology, universities, and schools of art and design. Some useful addresses are listed in Sect. 12.1. The excellent collections of color atlases and books on diverse aspects of color both at the Art and Architectural Library at Yale University (The Faber Birren Collection on Color), New Haven, Connecticut [8.39–41], and at the Royal College of Art (The Colour Reference Library), London, England [8.42], are well known. Most of the atlases listed in Table 8.1 are among the more than 60 listed (1982) in the Faber Birren Collection.

8.4 The Munsell Color System

The most important color-sample system currently used in the United States is the Munsell color system (Sect. 1.1). The Munsell notation has been incorporated in the Standards of the American National Standards Institute and the American Society for Testing Materials [8.19,19a]. The Japanese color standards are based on the Munsell notation [8.4] and the British Standards Institute uses it in its designation of standard paints [1.12].

The Munsell color system offers two collections of standard painted samples: a matt-finish collection (about 1325 color chips) and a glossy-finish collection (about 1600 color chips). Both collections are increased from time to time, whenever more saturated pigments of acceptable permanence become available. The standard color samples appear in chip form in the *Munsell Book of Color* (two volumes) [1.34]. The samples are also available as cards in file boxes and as loose sheets. Inexpensive, small-sample student sets (matt finish), of less than standards quality, are available for color instruction.

In this system, surface colors are identified by three quantities: *Munsell Hue*, *Munsell Chroma*, and *Munsell Value*. (Note that in this book the three Munsell terms are written with the first letter capitalized.) They permit quantitative specification of surface colors under specified conditions of viewing: average daylight (CIE ILL C), 45° illumination, and viewing along a sight line perpendicular to the surface [Ref. 7.2, p. 7]. A neutral gray background is usually used when the color of a sample is identified by comparing it with Munsell color chips. Specific recommendations are given in [Ref. 7.2, p. 7] for variations of the technique for determining the colors of matt and

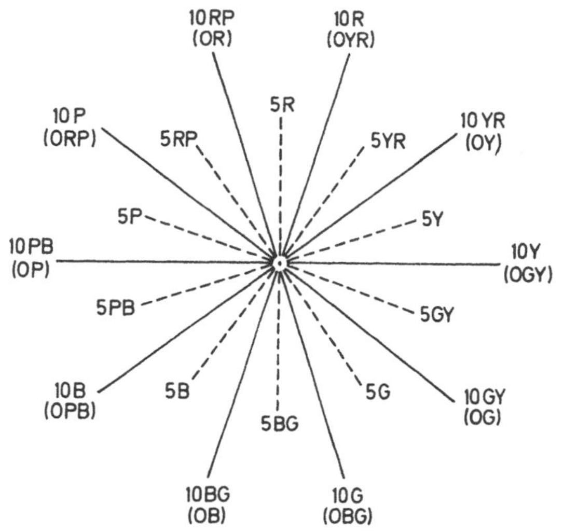

Fig. 8.9. The 10 Hue ranges of the Munsell Hue circle

glossy surfaces, satin-finished textiles, liquids, glasses, fluorescent materials, microscopic specimens, etc.

There are 10 *Hue ranges* in the *Hue circle* of the Munsell system, which appear in the order (clockwise) (Fig. 8.9): R (red), YR (yellow red), Y(yellow), GY (green yellow), G (green), BG (blue green), B (blue), PB (purple blue), P (purple), and RP (red purple). The Hue circle is subdivided by a scale consisting of 100 equally spaced *Hue radii*. A Hue range (for example R) includes eleven Hue radii, 0–10; the terminal Hue radius 10 of one range coincides with the initial Hue radius 0 of the next range. For each Hue range, there is a *major Hue,* which is located at the middle of each Hue range – that is, along Hue radius 5. The major Hues are designated 5R, 5YR, 5Y, 5GY, and so on. The Hues along the terminal radii of the ranges are designated 10R, 10YR, 10Y, 10GY, ... Figure 8.9 shows the radii for the major Hues (dashed lines) and the radii for the terminal Hues (solid lines). The numbering of the radii progresses clockwise from 0 to 10 in each range. Because the Hue along each terminal radius is identical with the Hue at the beginning of the next range, Hue 10R, for example, is identical with Hue 0YR. But the designation 0YR is not customarily used. This is similar to the hour of the day given in schedules for train and air travel. The end of the day, midnight, is given by 24:00. That moment may also be given by 0:00, the beginning of the next day, but the designation 0:00 is not used. But three minutes after midnight is indicated 0:03, and similarly on the Munsell Hue circle a Hue that is a bit more yellow than 10R might be, for example, 0.2YR.

Munsell color chips are provided not only for Hues at radii 5 and 10 in each of the ten Hue rangs but also for Hues at intermediate radii 2.5 and 7.5. Thus, the collection of the chips provides for a total of forty Hues: 2.5R,

5R, 7.5R, 10R; 2.5YR, 5YR, 7.5YR, 10YR; 2.5Y, 5Y, 7.5Y, 10Y; and so on for the seven remaining Hue ranges. The equal angular spacing ($9°$) of the forty Hue radii nominally represents the Hue spacing of the samples, which are approximately perceptually equal-spaced [Note 8.1].

In the collection of glossy color chips, supplementary colors of intermediate Hue are provided for Chromas of 12; 12, 14; and 12, 14, 16 at one or more Value levels in the range $V = 3$ to 8.5. Four intermediate Hues are provided for the R range: 1.25R, 3.75R, 6.25R, and 8.75R (and four Hues similarly for the YR, Y, and RP ranges); three Hues for the PB range: 3.75PB, 6.25PB, and 8.75PB; two Hues for the GY range: 1.25GY, and 8.75GY; and one Hue for the G range: 1.25G.

Munsell Value is designated on a scale from 0 to 10. The Munsell Value of a color is an indication of the lightness of perceived color, much as the luminance factor is. Munsell Value may be determined from the luminance factor by calculation or, more conveniently, by reference to Table 8.2 or to the more detailed Table 12.1 in Sect. 12.2. Munsell color samples are offered at Values 2, 3, ..., 9 for all Hues and also at Value 8.5 for the yellow Hues (only).

Munsell Chroma is often considered to be the approximate psychophysical counterpart of perceived saturation. The Munsell Chroma of a color sample is defined as its difference from neutral gray of the same Value. The Chroma scale is measured along a Hue radius: Chroma is zero at the center (neutral gray) and increases outward progressively to a maximum Chroma at the MacAdam limit determined for each Hue and Value. (The maximum Chromas are tabulated in [8.43].) Munsell color samples are offered at Chroma 1 for alternate Hues (..., 5R, 10R, 5YR, ...) and at Chromas 2, 4, 6, ... up to the maximum producible with pigments of acceptable permanency for each of the 40 Hues. (Some glossy color chips in the high Chroma range are provided for several intermediate Hues, as mentioned above.) The uniform steps of Chroma (2, 4, 6, ...) nominally represent the Chroma steps of the samples, steps that are approximately perceptually equal [Note 8.1].

A *Munsell Hue-Chroma diagram* (Fig. 8.10) is obtained when, for a given Value V, lines of constant Chroma (concentric circles) are superimposed upon a symmetrical pattern of Hue radii. The diagram shown illustrates by means of black dots all colors at Value 5 for which glossy Munsell samples are presently available. The central point (zero Chroma) represents a neutral gray sample. The Chroma circles display uniform steps at two-Chroma-unit intervals from Chroma 2 to 16.

Dominant wavelength and purity are not accurate indicators of perceived hue and saturation (Sect. 6.4). Judd wrote, "... Munsell hue, value, and chroma reflect the psychological facts of object color to a good approximation, whereas dominant wavelength, luminous directional reflectance [luminance factor] and excitation purity reflect them only to a poor approximation" [Ref. 6.7, p. 852].

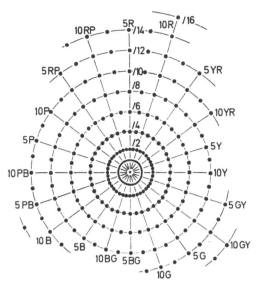

Fig. 8.10. Munsell Hue-Chroma diagram showing available Munsell standard color chips. The chips are indicated by dots that represent colors of 40 Hues and Chromas (up to 16) at Munsell Value 5 (luminance factor $Y = 0.20$). The Hue radii are not shown for colors of intermediate Hue: 2.5R, 7.5R, 2.5YR, etc.

It should be noted that, although the Munsell color system specifies uniform measures of Value and Chroma, the units of one are not equal to those of the other. This is shown by the fact that 1 Value unit equals 10 NBS units (Sect. 7.10), but 1 Chroma unit equals 7 NBS units. Thus, 1 Value unit is equivalent to about 1.5 Chroma units. A Munsell Hue unit is an angle measure (3.6°), 1/100th of the Hue circle (Fig. 8.11). The consecutive series of units of the Hue scale in a Hue circle of constant Chroma is intended to represent equal measures or steps of perceived Hue difference. These equal steps increase in magnitude as the radius (or Chroma) of the Hue circle increases (Fig. 8.10). Specifically, 1 Hue unit equals 0.4 NBS units at Chroma 1 and 3.3 NBS units at Chroma 5. This indicates that 1 Value unit equals about 25 Hue units at Chroma 1 and 3 Hue units at Chroma 5. [Ref. 1.18, p. 317; 3.13; 6.2, p. 175]

Plate VII shows black dots that represent presently available Munsell glossy chips of maximum Chroma for 40 Hues at Value 6. The heavy line drawn through the dots encloses all available Munsell glossy samples at Value 6. The line is reproduced in Fig. 8.12 on a reduced scale to permit comparison of the gamut with the larger gamut theoretically possible (MacAdam limit) at Value 6. A great difference is indicated between the present maximum (Chroma 10) for glossy chips of Hue 5G and the limiting Chroma 28. Perhaps this difference will be decreased somewhat by the introduction of new stable pigments.

Some standard Munsell and commercial color samples are also displayed in Plate VII. The colors of the Munsell samples have Munsell Value 6 ($Y = 0.30$); for the other samples, $V = 6$, approximately (Table 7.2).

117

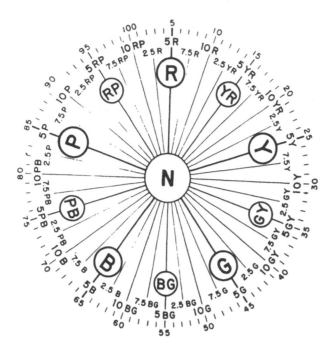

Fig. 8.11. Munsell Hue circle showing three ways to designate Munsell Hue: Hue numbers (*outer circle*); numbers and letters (*intermediate circle*); letters (*inner circle*). (Courtesy of Munsell Color, Baltimore)

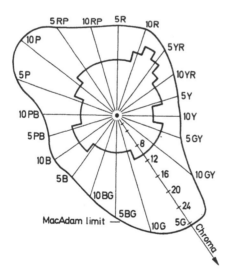

Fig. 8.12. Gamut of colors of available Munsell standard color chips reproduced on a smaller scale from Plate VII for comparison with the MacAdam limit at Munsell Value 6 (luminance factor $Y = 0.30$) (CIE ILL C)

Two of the commercial samples fall outside the gamut of the Munsell samples. Plates II and VII may be compared to judge their relative merits in displaying color samples (Sect. 7.5).

The painted chips in the *Munsell Book of Color* are grouped so that only one Munsell Hue, say 5YR, is represented on a page; the chips are arranged on a square grid so as to display variations of Munsell Value vertically and

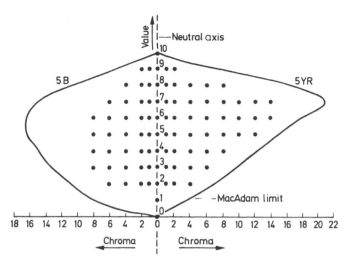

Fig. 8.13. Gamut of colors (opposing Hues 5B and 5YR) of available Munsell standard color chips indicated for comparison with the MacAdam limits (CIE ILL C)

Munsell Chroma horizontally. Figure 8.13 illustrates the same arrangement in a vertical plane in Munsell color space (discussed below) for two opposing Hues 5YR and 5B. Chroma is shown to increase radially from the vertical *neutral axis*. The neutral grays are represented on the neutral axis over the range from Value 0 (black) to Value 10 (white). For Hue 5YR and Chroma 4, for example, seven chips are indicated that vary in lightness from Value 2 up to Value 8.

Also shown in Fig. 8.13 are the MacAdam limits for Hues 5YR and 5B [8.43]. In the case of yellow red YR, chips are available up to Chroma 14, which approach the MacAdam limit rather closely. In the case of blue 5B, there is a greater gap between what is available in Munsell chips and what is theoretically possible.

Figure 8.14 shows two disks, one directly above the other in space. We can imagine such disks at equally spaced levels, represented by Values 1 to 9; each disk contains a circular array of points or dots, such as those shown in Fig. 8.10. Then, if the imagined disks are made to disappear, the arrays of dots remain in space, flat clouds of dots in *Munsell color space*. Each dot represents a different Munsell color (there are about 1600 glossy color chips!). The part of Munsell color space occupied by all of the clouds of dots has the form, roughly, of an onion; this portion of color space is called the *Munsell color solid*. Munsell color space, on the other hand, is bounded by the MacAdam limits; it has, let us say, the form of a turnip. (A model of Munsell color space is shown in [8.43].) We can therefore imagine the color solid within color space as an onion positioned within the shell of a large turnip. A vertical cross section of such a combination is shown in

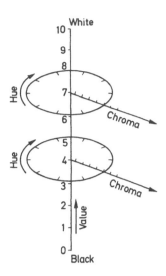

Fig. 8.14. Cylindrical arrangement of Hue, Chroma, and Value in Munsell color space

Fig. 8.13 where the MacAdam limit provides a profile of the turnip and the dots suggest the onion (deformed!). Similarly, a horizontal cross section is shown in Fig. 8.12.

A three-dimensional display of glossy color chips in the Munsell color solid is presented in Plate VIII. Only radial planes of the ten major hues are included in this model.

Munsell notation is easily described by an example: a yellow chip, designated by Hue 7.5Y, Value 7, and Chroma 8. In Munsell notation, the Chroma is written /8 as in Fig. 8.10, and the color is designated by 7.5Y 7/8. The neutral grays are indicated by the letter N because no Hue designation would be appropriate. Although the neutral grays have zero Chroma, 0 is not written in the notation. A neutral gray of Value 6 is designated simply by N 6/.

In the above paragraphs, Munsell Hue is represented by a notation consisting of numbers and letters (for example, 7.5BG). This number-letter notation is used in a Hue scale that is very widely employed and is utilized in the labelling of Munsell color chips. There are two other types of Hue notation, which are used much less frequently. Figure 8.11 shows three circles, each contains one series of Munsell Hue symbols. The number-letter series (..., 5RP, 7.5RP, 10RP, 2.5R, 5R, ...) is shown in the intermediate circle. In the outer circle the Munsell Hue-number notation (..., 95, 96, 97, 98, 99, 100, 1, 2, 3, 4, ...) is given. It passes in equal angular steps (3.6°) from 0 to 100 in the clockwise direction. At any given level of Chroma, the number scale divides the Munsell Hue circle into 100 equal units [Ref. 1.18, p. 381]. The numerical scale is precise and convenient for purposes of interpolation or judging intermediate Hues, but a number does not bring a Hue to mind as easily as a number-letter designation does. The third scale, indicated in the inner circle, possesses a letter notation spanning the Hue circle in 40 steps (..., RP, rRP, RP-R, pR, R, yR, R-YR, rYR, YR, ...). Only the letter symbols for the major Hues are shown in Fig. 8.11. Symbol N at the center represents a neutral gray, white, or black. Figure 8.11 is particularly useful for direct conversions between the 100-unit Hue scale and the number-letter Hue scale.

Color specifications given as $CIE(x, y, Y)$, CIE ILL C, can be converted to Munsell notation by the use of Table 12.1 and the set of nine charts [3.14, 8.20] presented in the Appendix (Figs. 12.1–9). The nine charts and five supplementary charts showing in greater detail the chromaticity range for pastel or grayish colors at Values 5 to 9 are available in large format (56 × 66 cm) [8.44]; a similar set in small format is found in [8.19a] [Note 12.1]. Table 12.1, which facilitates conversions between Munsell Value and luminance factor Y, is also presented in abbreviated form as part of Table 8.2. Conversions to Munsell notations usually introduce fractional numbers for Hue, Value, and Chroma. For example, a designation might be 8.4 Y $7.3_6/8.9$.

The 1929 edition of the *Munsell Book of Color* was for many years the authoritative source for Munsell notation. Notations determined with its use were called *Munsell Book notations* [Ref. 7.2, p. A-1]. In a report published in 1943, a committee of the Optical Society of America improved the spacing of the samples and extended the Munsell notation to the MacAdam limits [8.20]. For some years, colors brought into conformity with the 1943 report were specified by what were called *Munsell renotations*. Now that a sufficient number of years have passed, so that there need be no confusion with the old Munsell Book notations, the term *Munsell notation* is used for new chips. The current editions of the *Munsell Book of Color* conform fully with the 1943 report.

At this point, two matters concerned with color perception should be mentioned briefly. Formerly, uniform steps of Value in the *Munsell Book of Color* were established by visual means. But, because observers often fail to agree in comparing the lightness of color samples at high Chroma [Ref. 2.5, p. 166], an OSA committee decided to define Value by a mathematical formula that relates it to luminance factor, which is based on data that can be measured accurately (cf. Table 12.1). Nevertheless, Munsell Value defined precisely in this way does not represent perceived lightness accurately. For example, at a given level of Value, the perceived lightness increases as the Chroma increases [3.12,13].

Another comment concerns Chroma directly. It was mentioned earlier that Munsell Chroma is usually considered to be the approximate counterpart of perceived saturation. Until recently, this notion was, it seems, not questioned. But now qualification of this concept must be considered, because experimental work by Evans has showed that saturation and brilliance are both combined in the perception of Chroma [Ref. 2.5, p. 168].

Table 8.2. Conversions between Munsell Value V, luminance factor (or luminous transmittance) Y [8.20], and metric lightness L^* [Note 7.3]. (See also Table 12.1)

V	Y	L^*	V	Y	L^*
0	0	0	**6.00**	**0.300**	61.7
0.510	0.00593	**5.00**	6.16	0.320	63.3
0.850	0.0100	8.99	6.33	0.340	**65.0**
0.941	0.0113	**10.0**	6.48	0.360	66.5
1.00	0.0121	10.6	**6.50**	0.362	66.7
1.44	0.0191	**15.0**	6.64	0.380	68.0
1.49	0.0200	15.5	6.78	**0.400**	69.5
1.50	0.0202	15.6	6.83	0.407	**70.0**
1.95	0.0299	**20.0**	**7.00**	0.431	71.6
1.95	0.0300	20.0	7.06	0.440	72.2
2.00	0.0313	20.6	7.33	0.480	74.8
2.31	0.0400	23.7	7.35	0.483	**75.0**
2.44	0.0442	**25.0**	7.46	**0.500**	76.1
2.50	0.0461	25.6	**7.50**	0.507	76.5
2.61	0.0500	26.7	7.58	0.520	77.3
2.87	0.0600	29.4	7.82	0.560	79.6
2.93	0.0624	**30.0**	7.86	0.567	**80.0**
3.00	0.0656	30.8	**8.00**	0.591	81.3
3.10	0.0700	31.8	8.05	**0.600**	81.8
3.31	0.0800	34.0	8.27	0.640	84.0
3.41	0.0850	**35.0**	8.38	0.660	**85.0**
3.50	0.0900	36.0	8.48	0.680	86.0
3.68	**0.100**	37.8	**8.50**	0.684	86.2
3.89	0.113	**40.0**	8.58	**0.700**	87.0
4.00	0.120	41.2	8.68	0.720	88.0
4.29	0.140	44.2	8.87	0.760	89.9
4.36	0.145	**45.0**	8.89	0.763	**90.0**
4.50	0.156	46.4	**9.00**	0.787	91.1
4.55	0.160	47.0	9.06	**0.800**	91.7
4.80	0.180	49.5	9.24	0.840	93.5
4.85	0.184	**50.0**	9.40	0.876	**95.0**
5.00	0.198	51.6	9.41	0.880	95.2
5.03	**0.200**	51.8	**9.50**	**0.900**	96.0
5.24	0.220	54.0	9.66	0.940	97.6
5.33	0.229	**55.0**	9.82	0.980	99.2
5.44	0.240	56.1	9.90	**1.000**	**100.0**
5.50	0.246	56.7	**10.00**	1.026	
5.64	0.260	58.0			
5.82	0.280	59.9			
5.83	0.281	**60.0**			

Finally, attention should be drawn to the *Munsell Limit Color Cascade* introduced in 1974 [8.45]. The Color Cascade consists of 24 cards (10 × 26 cm); each of which contains two series of 16 colors of approximately constant hue and of varying Munsell Value and Chroma. The 768 colors are printed with glossy inks; each colored area measures 4.4 × 3.2 cm. The set includes colors of high Chroma that are outside the gamut of glossy colors of the Munsell collection.

Table 8.3. Munsell notation for Color Cascade hue series 11 and 12. Colors are identified by Cascade hue H (rows) and Cascade lightness L (columns). (CIE ILL C. Surface reflection excluded)

	1	2	3	4	L 5	6	7	8
$H\,11$	9.1BG 8.7/1.5	3.0B 8.1/3.6	3.4B 7.4/6.4	3.3B 7.0/8.0	3.3B 6.6/9.2	3.0B 6.0/10.6	3.2B 5.5/11.1	3.4B 5.2/11.7
$H\,12$	6.8BG 8.8/1.4	0.5B 8.2/3.6	0.8B 7.5/6.6	0.3B 7.1/8.0	0.3B 6.7/9.0	10BG 6.3/10.2	9.9BG 5.7/10.8	9.8BG 5.4/11.4

	9	10	11	12	L 13	14	15	16
$H\,11$	3.6B 4.6/10.7	3.8B 3.8/9.8	3.6B 3.2/8.5	3.6B 2.8/7.3	3.4B 2.5/6.4	3.0B 2.1/5.2	2.0B 1.7/3.2	0.2B 1.4/1.7
$H\,12$	9.4BG 4.8/10.6	8.8BG 3.9/10.0	8.4BG 3.5/9.1	7.8BG 3.1/8.1	7.4BG 2.9/7.2	7.1BG 2.6/5.7	5.8BG 2.1/4.0	4.2BG 1.6/2.3

The colors are identified with a simple notation. For example, 11–8 designates a blue of Cascade hue number 11 and Cascade lightness 8. Chroma is not indicated; for each color, the Chroma is the maximum available for the colorants employed. The Cascade lightness range for each of the 48 hue series is divided into 16 steps. The palest colors are found at lightness level 1 and the darkest at level 16. In general, the most vivid colors occur at level 8. The word "cascade" reflects the appearance of a hue series: From lightness level 8 (or in some cases 9) the Chroma of the colors "fall off" in the steps of the seven tints toward white and in the eight shades toward black.

The set includes tables giving luminance factors with respect to CIE ILL A and to CIE ILL C for each color. Another tabulation presents the Munsell notations of the colors. The Munsell notations for two Cascade hue series (11 and 12) are given here in Table 8.3.

The color cards are clearly most useful in applications that involve surface colors of maximum Chroma at any lightness level. They are proposed for color specification in varied domains, for example, in architecture, printing, and photography. A photograph of a card (reduced to 60 % of full size) showing blue and blue green colors (of Cascade hue series 11 and 12) at lightness levels 1 to 8 is presented in Plate IX (Fig. 11.11 and Sect. 11.10).

8.5 The Color Harmony Manual and the Ostwald Color System

The *Color Harmony Manual*, produced "chiefly to promote the knowledge and study of color harmony and color coordination in design" [Ref. 1.18,

p. 251], is based on a color system devised by the chemist (and amateur painter) Wilhelm Ostwald (1853–1932) [1.17, 8.11]. The colors of the color chips contained in the four editions of the *Manual* correspond to specific points within the Ostwald color solid (but with certain necessary modifications of the theoretical requirements and with certain accommodations made in the third edition for additional color chips). But before discussing the *Manual* in some detail, let us first review salient features of the Ostwald Color System.

In developing his color system, Ostwald confined his attention to surface colors perceived under nonisolated conditions (*related colors*). This permitted him to describe a color by perceived hue, whiteness, and blackness. [In the case of isolated (*unrelated*) colors, blacks and grays are not perceived.] He assumed, not entirely correctly, that all related colors can be designated by additive mixture of full colors, white, and black [Ref. 7.44, p. 147]. *Full colors* (also called *semichromes*) are ideal colors that contain no whiteness and blackness [Ref. 1.18, p. 379]. A full color is produced by an illuminated surface characterized by one of four types of ideal spectral reflectance curves [8.12].

Ostwald developed initially a rather extensive color system. What is discussed here is his subsequent abridgment of that system [Ref. 1.15, p. 64].

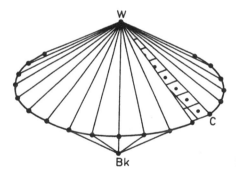

Fig. 8.15. Ostwald color solid (double cone)

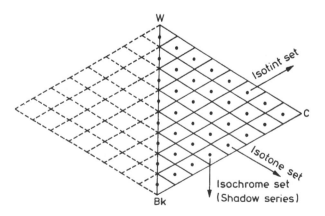

Fig. 8.16. Vertical cross section of the Ostwald color solid

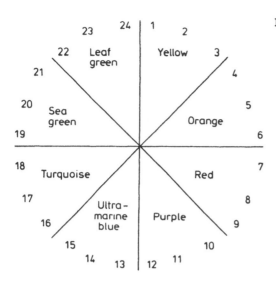

Fig. 8.17. Ostwald hue circle

(The abridged version is what the *Color Harmony Manual* is based on.) The color solid of the abridged system is shown in Fig. 8.15, and a vertical cross section of the color solid is given in Fig. 8.16. The Ostwald color solid is a *double cone*, consisting of two identical cones that have a common circular base and a central axis oriented vertically.

The cross section (Fig. 8.16) illustrates two of the 24 triangles radiating from the central axis. Each of the triangles represents a set of 28 colors of one hue. (Ostwald called them "monochromatic triangles" [Ref. 1.15, p. 92].) The 24 *Ostwald hues* are labelled by *Ostwald hue number* and the arrangement of the hue triangles in the color solid is indicated (top view) by the Ostwald hue circle shown in Fig. 8.17.

The vertical axis in the abridged color system is composed of his "practical gray series", extending in eight perceptually equal steps from a practical near-white W down to a practical near-black Bk ("printer's black"). The six intermediate neutral grays can be reproduced on a spinning disk (Sect. 5.8) divided into two sectors, one pure white and the other pure black. The percentage of the disk's area occupied by white is taken as the percentage of white in the resulting neutral gray. The percentages of white and of black for producing the near-white a, the near-black p, and the intermediate neutral grays ($c, e, g, i, l,$ and n) are given in Table 8.4.

Given the structure of the W-Bk axis, we can develop the notation and the compositions of the chromatic colors on each of the 24 hue triangles radiating from the W-Bk axis. The two-letter notation assigned to each point (Figs. 8.16,18) is determined by reference to the two diagonal paths that intersect at the point and originate at the W-Bk axis (Fig. 8.19). Thus the notations ic is determined by the diagonal path that rises from i on the W-Bk axis and by the diagonal path that descends from c. In each

125

Table 8.4. Percentages of pure white and pure black in the Ostwald neutral grays [Ref. 1.15, p. 64]

Neutral gray	a(W)	c	e	g	i	l	n	p(Bk)
Pure white	89.0	56.0	35.0	22.0	14.0	8.9	5.6	3.5
Pure black	11.0	44.0	65.0	78.0	86.0	91.1	94.4	96.5

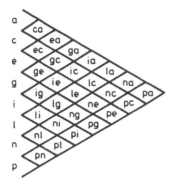

Fig. 8.18. Ostwald's two-letter notation

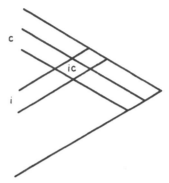

Fig. 8.19. Development of Ostwald's two-letter notation from the symbols for neutral grays

two-letter notation, the *first letter* is that on the *rising* diagonal path. A complete Ostwald color specification consists of the Ostwald hue number and the two-letter notation, for example, 14*ic* for a blue color.

The composition of an Ostwald color of any hue, for example color 14*ic*, can be determined with reference to Table 8.4. The diagonal that rises from *i* is a line of constant white content; it is called an *isotint line;* all Ostwald colors along it form an *isotint set.* The white content of color 14*ic* and all members of the set, including the neutral gray *i*, is 14 %. Similarly, the diagonal that descends from neutral gray *c* is a line of constant black content, an *isotone line.* All members of the *isotone set* that includes *ic* have a black content of 44.0 %. The full-color content at color 14*ic* is determined by subtracting from 100 the sum of the white and black contents (percentages); thus the full-color content is 42 %.

On each hue triangle (as indicated above) the color composition at *ic* is the same: 14 % white, 44 % black and 42 % full color. This composition indicates that a color at *ic* can be produced on a spinning disk having sectors of 14 %, 44 %, and 42 % area colored in pure white, pure black, and full color, respectively. The color of highest full-color content (color C) is located at the apex *pa* of each hue triangle; its composition is given by 3.5 %

white, 11.0 % black, and 85.5 % full color. The composition of near-white W is 89.0 % white and 11.0 % black; that of near-black Bk is 3.5 % white and 96.5 % black.

In addition to the isotint and isotone sets of colors, there are two other sets of interest in the Ostwald color solid. One, called the *isochrome set*, or more commonly, the *shadow series*, consists of colors represented by points falling along a straight vertical line (Fig. 8.16). There are six shadow series for each Ostwald hue. In each shadow series the ratio of the amount of full color to that of white, called *Ostwald purity*, is the same [Ref. 1.18, p. 379]. The colors in a shadow series decrease in lightness from the top point to the bottom one; both the full color and white contents decrease and the black content increases. The term "shadow series" is suggested by this variation of black content.

The remaining set, called an *isovalent circle*, consists of a circle of colors of all 24 hues that have the same percentage compositions of white, black, and full color. An isovalent circle is designated solely by a two-letter color notation, for example, *ic*. The isovalent circle of colors of maximum purity *pa* lies along the equator of the Ostwald color solid (double cone).

The arrangement of colors in the *Color Harmony Manual* follows in general that employed in the Ostwald System. Twenty-eight color chips of approximately the same hue (as measured by dominant or complementary wavelength, CIE ILL C) are displayed on sheets in a triangular arrangement. These hue triangles are presented in 12 pairs such that each pair consists of two complementary (additive) hues. At the apex of each hue triangle, the letter C refers to the color of maximum purity of the particular hue. One shadow series (indicated by a vertical set of points in Fig. 8.16) represents in CIE terms colors of approximately the same dominant or complementary wavelength and the same purity (hence approximately the same chromaticity).

Each color chip in the *Manual* is identified in the Ostwald manner – that is, by Ostwald hue number (designating the hue triangle) and an Ostwald two-letter notation (designating the location of the color in the triangle but, see below, not indicating the composition of the color). Table 8.5 presents for each Ostwald hue number the measured dominant or complementary wavelength (with respect to CIE ILL C) and purity of the colors of the matt color chips (first edition of the *Manual*) selected for the *pa* position at each triangle's apex (color C).

The first (1942) and second (1946) editions of the *Color Harmony Manual*, by E. Jacobson, contain 680 square color chips – namely, 28 color chips on each of the 24 hue triangles and eight chips for the W–Bk axis [8.10]. (The collection of color chips was developed by C.E. Foss.) In the first edition, the chips measure 1.6 × 1.6 cm; in the second edition, 2.5 × 2.5 cm. The chips were prepared by applying paint to one side of a transparent cellulose acetate film so as to present a matt surface on one side and a glossy surface

Table 8.5. Color specification of the 24 apex colors *pa* in the *Color Harmony Manual*, first edition [8.10]. Dominant (λ_D) or complementary (λ_C) wavelength, purity (p_e), and luminance factor (Y), with respect to CIE ILL C, are given for *pa* at each hue number

Hue No.	p_e [%]	λ_D, λ_C [nm]	Y	Hue No.	p_e [%]	λ_D, λ_C [nm]	Y
1	83.2	574.4	0.7513	13	66.0	469.4	0.0939
2	87.4	577.6	0.6945	14	57.5	475.4	0.0897
3	86.4	583.4	0.5243	15	48.6	480.6	0.1073
4	83.4	588.3	0.4238	16	43.8	483.8	0.1170
5	80.6	594.3	0.3224	17	40.1	486.6	0.1289
6	78.1	601.7	0.2481	18	37.6	488.9	0.1430
7	65.4	612.0	0.1300	19	32.5	491.9	0.1520
8	42.4	493.7c	0.1095	20	29.9	494.0	0.1618
9	41.0	502.0c	0.1022	21	23.2	501.5	0.1928
10	36.9	432.0c	0.0930	22	24.3	526.5	0.2209
11	34.2	558.4c	0.0948	23	55.9	551.5	0.3120
12	34.4	566.0c	0.0901	24	74.4	564.1	0.5260

on the other. Both matt and glossy sides are identified by the same Ostwald hue and two-letter notation. The color chips in both editions may be removed from their mounts to make color comparisons.

The third (1948) edition of the *Manual*, by Jacobson, Granville, and Foss, includes 263 color chips in addition to the 680 already employed in the earlier editions [8.46]. The chips are hexagonal in shape and measure 2.2 cm between opposing sides. The additional color chips were provided to improve representation in commercially important color ranges [Ref. 1.18, p. 251]. Of particular interest are the six added intermediate hue triangles of which four form two complementary pairs (Ostwald hue numbers: $1\frac{1}{2}$, $13\frac{1}{2}$; $12\frac{1}{2}$, $24\frac{1}{2}$; $6\frac{1}{2}$; $7\frac{1}{2}$). Additional shadow series that each have seven colors were included for 12 hue triangles, and a "light" isotone set was introduced for four hues $(24\frac{1}{2}, 1, 1\frac{1}{2}, 2)$.

Foss, Nickerson, and Granville have analyzed the Ostwald Color System and have shown that not all physically possible surface colors, whose gamut is defined by the MacAdam limits, are included by the Ostwald color solid [8.12]. On the other hand, they found that color chips that correspond to some of the points designated on the hue triangles cannot be produced with available colorants.

In the *Manual*, the colors at the *pa* positions were chosen close to Ostwald's specified colors. To make this choice the theoretical *pa* compositions had to be ignored. But, given the *pa* colors, a new gamut established on the basis of Ostwald's scheme would still fall "in some regions ... considerably inside the range of available colorants. This restriction has been avoided by a modification which still preserves the principal Ostwald concepts" [8.10]. (Ostwald's proposed principal of color mixture by averaging, i.e., disk mixture, was not followed in establishing the series of compositions between *pa* and black and white; instead the series "were made to cover the maximum gamut possible with the pigmented coating used" [Ref. 1.18, p. 250].) The "principal Ostwald concepts" preserved are: constant chromaticity of the colors in each shadow series, complete triangular arrays of colors of constant hue (given by dominant or complementary wavelength), and opposing triangles of complementary (additive) hues. "The color chips in the handbook [*Color Harmony Manual*] are the result of due consideration of these requirements" [8.10].

The *Manual* has been of interest to artists, because the isotint and isotone sets offer rather good approximations to the color gradations observed in nature. A shadow series (isochrome set) simulates the perception of different levels of light and shade [Ref. 6.2, p. 171]. Thus, the shaded greens of a vertical green pole illuminated from one side may be represented in a painting by a shadow series of greens.

8.6 The German Standard Color Chart

The *German Standard Color Chart* (DIN-Farbenkarte, DIN-6164), which has certain similarities to the Munsell Book of Color (Sect. 8.4) and has definite roots in the $\text{CIE}(x, y, Y)$ system, is used extensively in color specification in the Federal Republic of Germany and elsewhere in central Europe.

Originally, the Standard offered samples in the form of gelatin filters. The series of filters later served as primary standards for the subsequent production of a series of painted color samples [8.13]. By 1962, a complete matt collection of 590 painted paper chips was available. The chips ($2.8 \times 2.2\,\text{cm}$) are displayed on 24 sheets (Series 1 to 24); each sheet corresponds to hues of one dominant or complementary wavelength, with respect to CIE ILL C [Ref. 1.18, p. 269]. A separate sheet (Series 25) contains 19 matt achromatic color chips. The chips are mounted in slots and may be removed from the sheets to make color comparisons. A collection of 1004 glossy color chips of the same size was completed in 1984. These are also presented on 24 sheets (Series 101–124), and the 19 glossy achromatic colors are included on an additional sheet (Series 125).

Tables are provided for the matt and glossy collections. They give conversions to both CIE 1931 (X, Y, Z) and CIE 1931 (x, y, Y) with respect to CIE ILL C. Munsell and Ostwald notations are tabulated for the matt collection [Ref. 3.14, p. 501]. In addition, individual sample cards that have $7.4 \times 10.5\,\text{cm}$ color areas are separately available for many glossy colors. A handbook version of the glossy collection has been published by DIN, entitled *Musterkarte zur DIN-Farbenkarte*. The 1004 colors are printed in ink on pages of the handbook.

In the DIN-6164 system, color is specified by three quantities: *DIN-Farbton T*, *DIN-Sättigung S*, and *DIN-Dunkelstufe D* [Refs. 1.18, p. 266; 5.3, p. 663; 8.13,47]. The DIN's English translations of these terms are *hue number T*, *saturation degree S*, and *darkness degree D*. Colors of constant T have the same dominant or complementary wavelength, with respect to CIE ILL C. "Farbton T is the dominant wavelength expressed on a perceptually equispaced scale" [8.13]. The 24 hue numbers are identified by color terms in Table 8.6.

In the Standard, all color chips of the same hue T are mounted on one sheet in a rectangular-grid arrangement. Thus, we find that the sheet labelled $T = 8$ contains bluish red chips of varying saturation degree S and darkness degree D. They are placed such that the lightest colors $(D = 1)$ extend along the top horizontal row, passing from $S = 1$ (at the left-hand border), through $S = 2$, $S = 3$, ... to the highest saturation degree. The second row accommodates chips of darkness degree $D = 2$, and so on for other rows to the bottom row of darkest colors $(D = 8)$. The gamut of color chips provided is limited by the availability of pigments of acceptable permanency. Indeed, the number of chips for one hue can differ significantly from the number for another. Glossy and matt color chips have been produced for $D = 1$ to 7, and in a few cases for $D = 8$. Blue purple $(T = 11)$ glossy chips are available for only $S = 1, 2, 3$, and 4. At the other extreme, bluish red $(T = 8)$ and yellowish green $(T = 22)$ glossy chips have been made for $S = 1, 2, \ldots, 10$. The series of neutral gray chips are provided for darkness degrees $D = 0.5, 1.0, 1.5, \ldots, 9.0, 9.5$.

Figure 8.20 shows the CIE 1931 (x, y) chromaticity diagram with radial lines of constant hue number T $(T = 1$ to 24$)$ passing from the center (CIE ILL C) to the spectrum locus and purple line. The fact that lines rep-

Table 8.6. Hues of the DIN-6164 color circle and their hue-number designations (Farbton T). Browns and brownish colors (high values of D, darkness degree) are indicated within parentheses

T	Hue description	T	Hue description
1	Greenish yellow (olive)	13	Violet
2	Orange yellow (olive brown)	14	Bluish violet
3	Yellow orange (yellow brown)	15	Violet blue
4	Yellowish orange (yellowish brown)	16	Reddish blue
5	Orange (brown)	17	Blue
6	Red orange (reddish brown)	18	Greenish blue
7	Red (red brown)	19	Blue green
8	Bluish red	20	Bluish green
9	Red purple	21	Green
10	Purple	22	Yellowish green
11	Blue purple	23	Yellow green
12	Red violet	24	Green yellow (olive green)

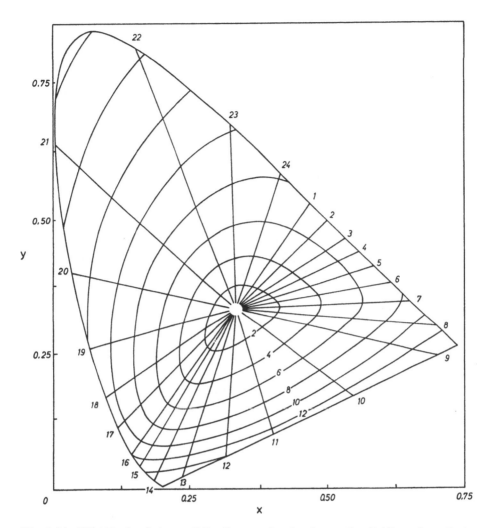

Fig. 8.20. CIE 1931 (x, y) chromaticity diagram showing the set of radial lines of constant hue number T and the set of curved lines of constant saturation degree S of the German DIN-6164 color system. Both sets of lines apply at all levels of darkness degree D. (From [Ref. 8.47, Fig. 30])

resenting constant perceived hue on a chromaticity diagram are, for the most part, curved (much as shown by the lines of constant Munsell Hue in Figs. 12.1–9) is a reminder that the straight radial lines of hue number T (like those of constant dominant or complementary wavelength) are only approximate representations of perceived hue.

Darkness degree D is a lightness quantity determined by calculation [Note 8.2]. The scale for D varies from 0 to 10. At zero, D represents maximum lightness (white); when $D = 10$, the lightness is zero (black). Darkness degree D for the color of a surface is related to a *relative luminance*, which is defined as the luminance factor Y divided by the maximum luminance factor (defined by the MacAdam limit) at the same chromaticity. It can be shown that, at constant Y, darkness degree D decreases as purity increases [3.13]. In general, there is no simple relation between perceived lightness and darkness degree D.

Saturation degree S, unlike purity (excitation purity) in the CIE (x, y, Y) system, is not related to the chromaticities of the colors of monochromatic lights. It is, however, based on data obtained in an experimental study in which it was established that a certain series of colors of a full range of perceived hues and of the same relative luminance produced the same perceived saturation. This initial series was designated $S = 6$ and is shown as a curve (oval) in Fig. 8.20. Similar series were established at other levels of saturation degree ($S = 1$–5, 7, 8). Because the spacing of the curves seemed to be uniform on the CIE 1960 (u, v) diagram, curves for constant values of S were defined by extrapolation for $S = 9$ and higher [8.13].

An outstanding feature (by definition) of the DIN-6164 system is that the oval patterns of S and the radial lines of T plotted on the CIE 1931 (x, y) chromaticity diagram are identical at all levels of D. Thus one diagram serves all levels of D. By contrast, the Chroma ovals and the Hue lines of the Munsell system plotted on the CIE 1931 (x, y) chromaticity diagram change when Value is varied (Figs. 12.1–9).

The form of the *DIN-6164 color solid* is basically that of an ice-cream cone (or a spherical sector) (Fig. 8.21). Hue numbers T are indicated around

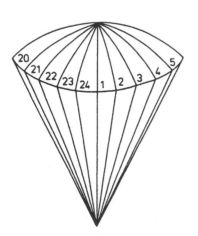

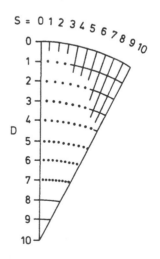

Fig. 8.21. The DIN-6164 color solid. Hue numbers T are indicated around the rim

Fig. 8.22. Vertical radial slice from the DIN-6164 color solid. The *black dots* represent available colors (glossy) of chips in Series No. 122 ($T = 22$, $D = 1$–7)

the circular rim. White is represented by the point at the top, and black by the point at the bottom of the cone. Figure 8.22 shows a vertical slice obtained along a plane of constant hue number $(T = 22)$ in the color solid. The radial slice shows a fan of straight lines of constant S and circular arcs of constant D. A line of constant S in the color solid represents colors that compose a shadow series, much as colors along vertical lines in the Ostwald color solid do (Sect. 8.5) [3.4]. For any given value of darkness degree D, the equal steps on the scale on the circular arc are intended to represent equal perceived differences in saturation degree S. It is clear in Fig. 8.22 that the size of the steps decreases as the darkness degree increases. The glossy chips available for hue $T = 22$ are indicated by dots in Fig. 8.22.

8.7 The Natural Colour System (NCS) and the Swedish Standards Color Atlas

The Swedish *Natural Colour System* (NCS) provides an effective means for everyone with normal color vision to make color evaluations without the use of color-measuring instruments or of color samples for comparison. The NCS can be employed directly for determining the perceived color of a wall in a room, of foliage in the distance, of a painted area in which simultaneous contrast occurs, of a spot on a television screen, etc. A color determined in this way is an absolute measure based on color perception. It differs often from psychophysical determinations, which rely on color matching. A. Hård and L. Sivik state that, although we may be able to distinguish between 10 million color stimuli under favorable conditions, "our ability to identify a color with some certainty is a great deal less," the total number of colors probably being about 10 000 or 20 000 [8.27]. They claim that this degree of precision can be met in the NCS.

The conception of the NCS can be traced to the German physiologist Ewald Hering (1834–1918), whose theory of color vision is the source from which much theoretical and experimental research in color perception has grown [2.3, 5.18]. The NCS was revived by the Swedish physicist Tryggve Johansson (1905–1960), and, since 1964, a program of research concerned with it has been pursued at the Scandinavian Colour Institute (formerly, the Swedish Colour Centre Foundation) in Stockholm [8.22–25, 48, 49].

Basic to the NCS is the recognition of the six *psychological primaries* (Sect. 3.2): white (W), black (S, for the Swedish word "svart"), yellow (Y), red (R), blue (B), and green (G). The last four are the *unitary hues:* yellow that is neither greenish nor reddish, red that is neither yellowish nor bluish, blue that is neither reddish nor greenish, and green that is neither bluish nor yellowish [Ref. 2.5, pp. 66, 107]. All other hues are recognized as mixtures of two unitary hues; for example, greenish yellows, reddish yellows, yellowish reds, bluish reds.

The first step in judging a color by the NCS is the determination of its hue [8.22, 24, 25]. This requires familiarity with the schematic Hering hue circle shown in Plate X [8.22]. (The sequence of hues is also shown in a straight band below the circle.) The binary compositions of hues, which fall between the unitary hues Y, R, B, and G, are represented schematically in the Hering circle. Of particular help is the NCS hue circle adapted from the Hering circle (Fig. 8.23), because it shows the hue scale (read clockwise) in NCS notation. In both circular diagrams, we can see that all positions in the quadrant of the hue circle between Y and R, for example, are occupied by a continuous gradation of binary mixtures of Y and R. At the midway position in the Y/R quadrant, the binary hue indicated by the radial dashed line in Fig. 8.23 and Plate X is given by the notation Y50R, representing a 50/50 mixture of unitary yellow and unitary red. (The binary hue notation is explained below.) Similarly, R50B, B50G, and G50Y represent 50/50 mixtures.

The dashed lines also demarcate hue ranges. Thus, the hues between G50Y and Y50R are the yellows. The yellows between G50Y and Y are greenish; those between Y and 50R are reddish. Continuing around the circle, we find yellowish reds and bluish reds; reddish blues and greenish blues; bluish greens and yellowish greens. In this terminology, common hue terms, such as orange, purple, and cyan are deliberately excluded. Browns and olives must be recognized as reddish and greenish yellows and yellowish greens when the luminance factor is low. The Hering hue circle indicates clearly that binary hue mixtures of red and green, or of blue and yellow, do not exist. Indeed we are unable to see any such hues as greenish reds, reddish greens, bluish yellows, or yellowish blues [Ref. 8.25, p. 111].

To judge hue, the observer must first identify the two unitary hues of which the hue is composed and the quadrant of the hue circle in which the hue is located. When this is done, the observer judges the relative proportions of the two unitary hues required to produce the hue. For example, the

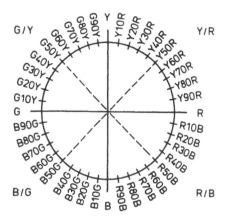

Fig. 8.23. NCS hue circle. The circle is divided into four quadrants (Y/R, R/B, B/G, G/Y) by the unitary hues Y, R, B, and G. The scale (read clockwise) shows standard NCS hue designations. (Based on [8.50])

observer may decide that the hue is a binary mixture of blue and green, requiring location in the B/G quadrant of the Hering circle. Imagine that, after some consideration, the hue is judged to be 70 % unitary green and 30 % unitary blue. This bluish green hue is located in the B/G quadrant, 70 % of the way along the arc (reading clockwise) from B to G (Fig. 8.23). The NCS notation for this hue is B70G, which means 70 % unitary green and the rest unitary blue. (The 30 % for unitary blue is not written, because it is easily obtained by subtracting the percentage for unitary green from 100.)

The next task is to judge the relative amounts of bluish green hue B70G (the chromatic component C), white W, and black S — that is, the *NCS chromaticness, NCS whiteness,* and *NCS* blackness, respectively. Let us say that the observer decides that the relative amounts are: S, 20 %; W, 30 %; and C, 50 %. Now all the information is available for the NCS specification of the color. It is: 2050-B70G. By convention, the relative amount of black S (20 %) is listed first, that of the chromatic component C (50 %) second, and finally the hue's specification (B70G). Only two of the relative amounts (S and C) are stated; there is no need to state the relative amount of the third (W), because it can be obtained simply by subtracting the sum of S and C from 100.

The NCS color solid is a double cone much like the Ostwald color solid (Fig. 8.15). Figure 8.24 shows a view of a double cone with indications for W and S at the north and south poles, lines W-Y, W-G-S, and W-B suggesting radial cuts by vertical radial planes of constant unitary hues Y, G, and B, and a line W-B70G-S for the radial plane of the binary hue B70G. The scale around the NCS hue circle in Fig. 8.23 may be imagined inscribed along the equator of the double cone. Each radial vertical plane cuts half-way through the color solid, ending at the vertical W-S axis; the result is a constant hue plane having a triangular shape. Figure 8.25 shows such a hue triangle for hue B70G.

The relative amounts of S, W, and C in a perceived color P are commonly represented by plotting the corresponding point for color P on the pertinent hue triangle. The color 2050-B70G is indicated by P on hue tri-

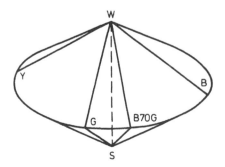

Fig. 8.24. NCS color solid (double cone). The straight lines on the surface of the double cone represent cuts made by vertical radial planes of constant hue that reach to the central W-S axis (*dashed line*). Those indicated are for planes of three unitary hues (B, G, and Y) and one binary hue B70G

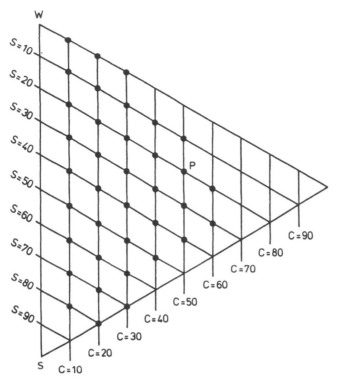

Fig. 8.25. NCS hue triangle. Vertical radial plane of constant hue in the color solid (double cone). The vertical lines are of constant NCS chromaticness and the slanted lines of constant NCS blackness. The *black dots* indicate chips for the hue B70G presented in the *SIS Colour Atlas NCS*. Point *P* at the intersection of lines $S = 20$ and $C = 50$ represents the color given by the NCS notation 2050-B70G

angle B70G in Fig. 8.25. The hue triangle shows slanted (descending) lines of constant NCS blackness ($S = 10, 20, \ldots, 90$) and vertical lines of constant NCS chromaticness [$C = 0$ (the W-S axis), 10, 20, $\ldots$, 90]. Point *P* is located by the intersection of lines $S = 20$ and $C = 50$. Slanted (ascending) lines of constant NCS whiteness have been omitted from the diagram because only two intersecting lines are required to locate point *P*. [It will be recalled that lines of constant z (Fig. 6.4) were omitted from Fig. 6.5 by similar reasoning.]

At the Scandinavian Colour Institute, extensive tests of the application of the NCS have shown that people with no particular knowledge about color and with no previous experience of color specification or color measurement are capable of making judgments of hue and the relative quantities C, W, and S unaided by color samples [Refs. 8.25, p. 116; 8.27] [Note 8.3].

The Swedish Standards Institution (SIS) has adopted the NCS as a Swedish standard for color notation and a color atlas (*SIS Colour Atlas NCS*, Swedish standard No. 01 91 02) as a practical illustration of the system

[8.23,27]. The *Atlas* displays 1412 matt color chips (1.5×1.3 cm). There are 40 full-page triangles, each showing color chips of constant hue. Twenty of the triangles display chips starting from $C = 10$; twenty alternate triangles show chips starting from $C = 40$. Figure 8.25 shows the triangle for hue B70G; the color chips provided in the *Atlas* are indicated by black dots. Unfilled positions in the triangle represent colors for which suitable pigments were not available. One full page from the *Atlas* showing a triangle that contains color chips of one hue has been published in [8.27]. A table is available that specifies the colors of all the chips by $\text{CIE}(x, y, Y)$.

It must be emphasized that the NCS is "a system for denoting colors as and when they appear to man" [8.27]. The *Atlas* is not required for making NCS color judgments. The *Atlas* is useful for the precise specification of colors on the basis of perceptual criteria, much as the *Munsell Book of Color* is. But the correct application of the *Atlas*, like that of the *Book of Color*, is restricted to specific conditions of illumination and background (white, NCS; middle gray, Munsell [3.13]).

The NCS and the Munsell system collections represent different samplings of essentially the same color space. For this reason, it has been thought that a simple relationship probably exists between them [8.26]. Such a relationship established by equations, charts, or tables would be of real interest to many because with their use NCS judgments of surface colors viewed in daylight could be translated into Munsell notation and then into $\text{CIE}(x, y, Y)$ notation without the use of color-measuring instruments [Note 8.4].

It is of interest here to recall that the hues of the colors of successive samples in a Munsell Hue circle were intended to differ by perceptually equal amounts [Note 8.1]. In the *Color Harmony Manual*, hues were selected so that opposing hues in the hue circle are complementary (additive), but perceptually equal hue steps were not obtained. In the NCS, the opponent unitary hues are diametrically positioned in the hue circle; with this arrangement perceptually equal hue steps are also not obtained. The inequality of the perceptual differences between the unitary hues is indicated in Table 11.1 and discussed in [Note 11.1], in Sect. 11.3, and in [Refs. 1.39, p. 60; 3.12,13; 8.26].

Although the color solids of the NCS and the *Color Harmony Manual* are both double cones, there is a major difference in the quantities represented on the hue triangles, i.e., slices from the color solid. For the *Manual*, the color samples were produced to match systematic variations of psychophysical quantities: purity and luminance factor [Refs. 1.18, pp. 250,284; 1.20, p. 165]. The *Manual* is based on stimulus variables [3.12]. On the other hand, in the NCS *Atlas* the samples were made to correspond to judgments of relative amounts C, W, and S – that is, judgments of perception. This perceptual basis is probably of significance to artists interested in the use of series of colors of constant NCS chromaticness (NCS shadow series) and of series of constant NCS blackness (NCS isotones) and of constant NCS whiteness (NCS isotints) (compare Figs. 8.16,25). Undoubtedly, the NCS *Atlas* will find many practical applications and may replace the Ostwald color system in certain areas of art and design.

8.8 Chroma Cosmos 5000 and Chromaton 707

Chroma Cosmos 5000 (1978) is a color-sample set developed by the Japan Color Research Institute (JCRI), Tokyo, for use in design [8.4–6]. The set contains 5000 glossy painted samples (1.1 × 2.2 cm) of colors that conform in order to the Munsell system. The samples are mounted on 23 double charts (folded). Unfolded they measure 26.7 × 74.9 cm. Each double chart displays two panels of samples of constant Munsell Chroma, arranged to show variations of Munsell Hue in horizontal rows and of Value in vertical columns. One double chart may contain as many as 350 different samples. Four double charts are shown in Plate XI.

The set includes all the colors presented in the *Munsell Book of Color* (glossy-finish collection), up to and including Chroma 14. It contains, in addition, many intermediate colors which results in smaller differences between adjacent colors. Consequently, there are about three times as many samples as are in the Munsell collection. Chroma is varied in unit intervals (1, 2, 3, ...) to the maximum available with present-day pigments of acceptable permanency, but the series is halted at Chroma 14. [In the Munsell collection, color samples of all Hues are represented by Chromas at two-unit intervals (2, 4, 6, ...) up to the highest available, which in some instances is Chroma 16.] Value is varied in 18 half-unit intervals (1, 1.5, 2, 2.5, ..., 9.5), whereas Values at unit intervals (2, 3, 4, ..., 9) are offered in the Munsell collection (Munsell yellows are offered also at Value 8.5).

At Chroma 1, *Chroma Cosmos 5000* (as in the Munsell collection) offers 20 alternate Hues – that is, Munsell Hues at 5-unit intervals (..., 5R, 10R, 5YR, 10YR, ...). At Chromas 2–10, it includes all 40 Hues of the Munsell series (..., 5R, 7.5R, 10R, 2.5YR, 5YR, ...) and, in addition, eight Hues are included from the supplementary series of Munsell samples of intermediate Hues (1.25R, 6.25R, 1.25YR, 3.75YR, 8.75YR, 6.25Y, 3.75PB, 6.25PB). At Chromas 11–14, it uses Munsell Hues 7.5P through 7.5Y and the six R, YR, and Y intermediate Hues given above [8.6].

The Munsell 40-Hue circle presents hues of colors of samples that are approximately perceptually equal-spaced. With eight Hues added to the 40-Hue circle, the particular feature of equal hue spacing is lost, but another is gained: The opposite Hues in the 48-Hue circle are approximately complementary, which is considered an important aid in design. Table 8.7 presents the opposing complementary pairs in the 48-Hue circle. The complementary pairs are not additive complementaries. They are approximate Munsell

Table 8.7. Complementary (PCCS) Munsell Hue pairs in the 48-Hue circle of *Chromo Cosmos 5000*

5R	2.5BG	6.25Y	10PB
6.25R	5BG	7.5Y	2.5P
7.5R	7.5BG	10Y	5P
10R	10BG	2.5GY	7.5P
1.25YR	2.5B	5GY	10P
2.5YR	5B	7.5GY	2.5RP
3.75YR	7.5B	10GY	5RP
5YR	10B	2.5G	7.5RP
7.5YR	2.5PB	5G	10RP
8.75YR	3.75PB	7.5G	1.25R
10YR	5PB	10G	2.5R
2.5Y	6.25PB	2.5BG	5R
5Y	7.5PB		

equivalents of complementary hues specified by the JCRI in their Practical Color Coordinate System (PCCS) (1965) proposed for purposes of design [8.4]. There is some rough resemblance between these complementary pairs and those found by Wilson and Brocklebank in their studies of afterimages (Sect. 11.7).

Each folded chart is accompanied by two transparent overlays on which are printed outlines that segregate underlying samples into groups or *color blocks* (there are 230 color blocks). Each block of colors is given a color-block name (for example, "light greenish yellow"). Figure 8.26 is a reproduction of a transparent overlay showing color blocks in the Hue range 10RP to 7.5YR at Chroma 6. This overlay is employed with one of the panels of color samples shown in Plate XI. Superimposed on the outlines of the color blocks shown are three curved lines that partition the array of colors into four *tone series:* light tone, soft tone, dull tone, and dark tone. (The PCCS employs the following tone groupings: vivid, strong; bright, light, pale; soft, dull; deep, dark; light grayish, grayish, dark grayish; white, light gray, dark gray, black.)

The PCCS specifies 12 pairs of complementary Hues, a lightness scale corresponding to that of Munsell Value, and a PCCS saturation scale. The PCCS saturation scale (0s, 1s, 2s, 3s, ..., 9s) was established by assigning the label 9s to the color that produced the highest saturation and by subdividing the range into perceptually equal steps. Munsell Chroma and PCCS saturation can be compared in Fig. 8.27, where alternate PCCS saturation lines (1s through 9s) are plotted on a Munsell Hue-Chroma diagram.

The PCCS was devised to serve as a "guide for harmonious expression of color combination". But the enormous project of producing the extensive set of samples was not undertaken [8.6]. It was decided instead to produce *Chroma Cosmos 5000* in its place by adapting the Munsell collection of samples to this purpose. "In *Chroma Cosmos 5000*, it is intended to interpret

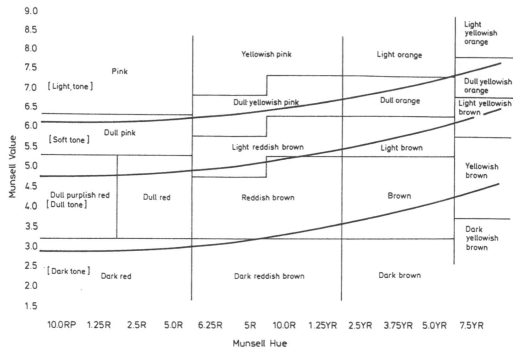

Fig. 8.26. Copy of a transparent overlay for one panel of a chart representing Munsell Chroma 6. It is the panel at the lower left-hand corner of Plate XI. (Courtesy of T. Hosono, Japan Color Research Institute, Tokyo)

the tone series indicated by PCCS and represented by Munsell color space" [8.7].

Munsell Chroma was adopted as an approximate equivalent of PCCS saturation [8.4], the Munsell Hue circle was increased to 48 Hues to provide opposing approximately complementary (PCCS) Hue pairs, colors of intermediate Chromas and Values were added to the Munsell set for improved precision, and transparent overlays were provided for use over the constant-Chroma charts to reveal the basic PCCS tone series in an acceptable way. The set of samples is accompanied by a text in Japanese and English. CIE(x, y, Y) equivalents are provided.

Chromaton 707 (1982) is a more recent development by JCRI [8.7,8]. It is a condensed, yet practical, version of *Chroma Cosmos 5000*. Its 707 colors have been selected from *Chroma Cosmos 5000*. Due consideration was given to their frequency of use in consumer products. The color chips are mounted on seven fold-out charts (26 × 45 cm), and, in addition, a complete set of swatches (4.5 × 12.7 cm) is supplied. The chips are mounted in a way

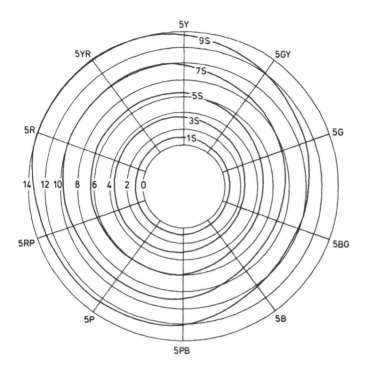

Fig. 8.27. Munsell Hue-Chroma diagram showing curves (*heavy lines*) that represent levels (1s, 3s, ..., 9s) of PCCS saturation for comparison with the concentric circles that represent levels (0, 2, ..., 14) of Munsell Chroma. The radial lines indicate constant Munsell Hues. (From [Refs. 8.4, Fig. 3; 8.7, Fig. 3])

that displays the PCCS tone series. Each chart presents 24 Hues (except one that presents 20) and a Value range of 17 half-unit intervals.

The seven charts are identified by tone as follows: (1) neutral and off-neutral tones (Munsell Chroma 0.5 to 1); (2,3) pale and grayish tones (Munsell Chroma 1 to 2, 2 to 3); (4,5) light, dull and dark tones (Munsell Chroma 3 to 5, 5 to 7); (6) bright, strong, deep tones (Munsell Chroma 8 to 11); (7) vivid and high-purity bright tones (Munsell Chroma 8.5 to 14).

9. Color Systems (Continued): The OSA Uniform Color Scales

9.1 Equally Spaced Colors

In 1947, when D.B. Judd at the U.S. National Bureau of Standards was asked by the National Research Council of the National Academy of Sciences whether he could suggest "some badly needed research program that would have nothing to do with armaments and war," he suggested a study of *uniform color scales* [8.31]. A uniform color scale is a linear sequence of colors that differ by perceptually equal amounts (Sect. 9.5). The Optical Society of America (OSA) decided to undertake the project, and its Committee on Uniform Scales was formed the same year with Judd as chairman. It turned out that years of difficult pioneering work were required, involving both the specification of the uniform sampling of color space and, finally, the production of precise color samples.

In 1977, the Committee announced the availability of the color samples, a collection of glossy acrylic paint color cards (5 × 5 cm) of 558 colors, 424 of which form an extensive set (referred to below as the "basic set") that represents a practical gamut of colors produced by present-day pigments of acceptable permanency. The extra 134 color samples ("intermediate samples") are included to be inserted midway between adjacent colors of the basic set in the central region of the gamut – that is, in the region of medium lightness, neutral grays, and low saturation. These extra colors serve to bisect color differences, creating thereby a second set, a half-step set, referred to below as the "pastel set" [8.31,32; 9.1,2].

The gamut of the colors is represented in OSA Committee color space by a specific geometrical arrangement of points. Any chosen point within the arrangement is surrounded by 12 nearest-neighbor points, and all nearest-neighbor points are at an *equal distance* from that point. Thus, for any color in the basic set (except those at the limits of the gamut), there are 12 neighbor colors that differ from it by the *same perceptual amount* (which is about 20 just-perceptible color differences [9.2]). In the pastel set, the arrangement is the same but the color difference is halved.

The selection of an arrangement involving 12 nearest neighbors met the OSA Committee's objective: the "provision of a set of colors with which the maximum number and variety of uniform color scales can be constructed" [9.3]. A model that consists of spheres painted in the 424 colors and mounted in accordance with the geometrical structure of OSA Committee color space (referred to below as "OSA color space") is shown in Plate XII. The pigments for the paints employed in preparing the OSA color cards are described in [5.11]; spectral reflectance curves for samples of paint films containing some of the pigments are represented in Figs. 5.5–9.

9.2 OSA Color Space

In order to select samples from the collection for assembling uniform color scales and color arrays, it is helpful to be familiar with the uniform geometrical structure of OSA color space. We can get a very good idea of it by considering a symmetrical cluster of 13 color-sample points in the color space: one color-sample point surrounded by 12 nearest-neighbor points, all at an equal distance [Refs. 1.39, p. 62; 8.32–34; 9.3–6].

The geometrical arrangement of the 13 points corresponds precisely to the points locating the 12 corners and the center of the geometrical form shown in Fig. 9.1 [9.4–6]. The form is called a *cubo-octahedron* (or *cuboctahedron*); it may be constructed by cutting off the eight corners of a cube. The cuts are made so as to leave eight identical equilateral triangular faces. Each vertex of each triangle touches a vertex of another. Between the triangular faces are six square faces, the remnants of the faces of the original cube.

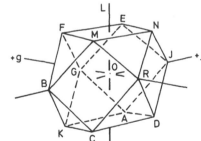

Fig. 9.1. Cubo-octahedron. The central point O and the 12 surrounding points at the corners (indicated) are all equidistant nearest-neighbor points

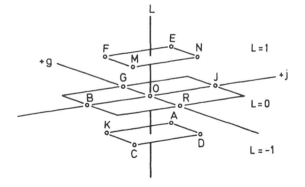

Fig. 9.2. The 13 equidistant points lie on square grids in three equally spaced parallel planes. In the orientation of the cubo-octahedron shown in Fig. 9.1, these planes are the OSA horizontal cleavage planes at lightness levels $L = -1, 0, 1$

143

It is important to note that each of the 12 points that surround point O has its own set of 12 nearest-neighbor equidistant points and that this three-dimensional lattice of points extends throughout OSA color space. For example, point N is also a central point with its nearest-neighbor points E, M, R, J, O, and seven other points in color space (not shown) (Fig. 9.1).

The second important feature is that the 13 points are located on square grids on equidistant horizontal planes (Fig. 9.2). Thus, in OSA color space all sample points are located at intersections on square grids on each of a set of equally separated horizontal planes. Furthermore, we should note that, although points E, F, M, and N (which define one square-grid unit in the horizontal plane at $L = 1$) are located directly above points A, K, C, and D in the plane at $L = -1$, they are displaced laterally by a half-grid-unit length with respect to points B, O, and R (on the intermediate plane at $L = 0$), which define a square-grid unit of the same size. Thus there is a regular alternation of the locations of the square grids on the successive horizontal planes.

We are now at a stage where we can consider how to designate a color in OSA color space by its location on a horizontal plane. The OSA method requires, however, the use of scales involving negative numbers. In this and in the following paragraphs I shall digress slightly to aid those for whom negative numbers may be an oddity, for, after all, negative numbers are not commonly experienced in daily life (the scale on a household thermometer being a notable exception). In the Munsell color system, negative numbers are not employed. A color is specified by Munsell Hue (Hue units 0 to 100), Munsell Value (0 to 10), and Munsell Chroma (0 to a maximum value), all in positive numbers.

In the OSA system, the coordinates of points are conveniently given with respect to the square-grid arrangement. As we shall now see, this also does not require the use of negative numbers [Note 9.1]. As in the Munsell system, let us assign positive numbers to successive horizontal planes, each of which cooresponds to a level of lightness, say Q. For the purpose of locating points in a square grid on a horizontal plane (at, say, lightness level $Q = 5$), we can introduce a pair of axes to serve as a reference, as shown in Fig. 9.3. As in a Munsell Hue-Chroma diagram (Fig. 8.10), which has a circular grid, we designate hue regions, such that the j axis extends into the zone exhibiting yellowness [to avoid the much-used letter y, j (for jaune, the French word for yellow) is employed in the OSA system] and the g, b, and r axes extend into the zones of greenness, blueness, and redness, respectively. (To conform with the OSA system, g is placed at the top; this is the inverse of what is found in the Munsell system.) In this arrangement, the central point ($j = g = r = b = 0$) corresponds to an achromatic color, i.e., no hue. Thus, in Fig. 9.3, color T is designated by lightness level $Q = 5$ and the grid coordinates $j = 4$ and $g = 2$. Similarly, colors U, V, and W are

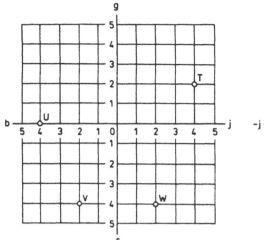

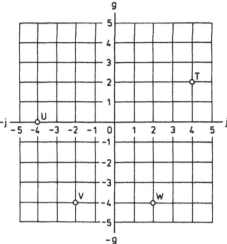

Fig. 9.3. Possible method (not proposed) of representing colors T, U, V, and W (of equal lightness Q) on a square grid without the use of negative numbers

Fig. 9.4. OSA method of representing colors T, U, V, and W (of equal lightness L) on a square grid

designated by lightness level $Q = 5$ and their coordinates; $b = 4$, $g = r = 0$; $b = 2$, $r = 4$, and $j = 2$, $r = 4$, respectively.

In the OSA system a horizontal plane cuts the vertical neutral axis at a medium gray of OSA lightness $L = 0$. Lighter colors are located at lightness levels above zero (positive numbers) and darker colors, at levels below zero (negative numbers), much like temperatures above the freezing point of water (0°C) being positive numbers on the Centigrade (Celsius) thermometer scale and those below 0°C being negative numbers.

Colors T, U, V, and W represented in Fig. 9.3 are treated again in Fig. 9.4. Here, in the OSA system, the j axis extends through 0 to negative numbers and the g axis extends through 0 to negative values. In this way, the notation has been simplified, replacing b by $-j$ and r by $-g$. Now T is located at lightness level L by $j = 4$, $g = 2$; U, by $j = -4$, $g = 0$; V, by $j = -2$, $g = -4$; and W, by $b = 2$, $g = -4$.

The form of the OSA notation employed is (L, j, g). Thus, if the lightness level L is -2, the colors T, U, V, and W can be specified simply by $(-2, 4, 2)$, $(-2, -4, 0)$, $(-2, -2, -4)$, and $(-2, 2, -4)$, respectively. The compactness of this notation is a merit. Although the previous system would have the advantage of suggesting hue (for example, the notation $b = 2$, $r = 4$ suggests a bluish red), with a little practice $-j$ can be remembered to indicate blueness and $-g$, to indicate redness. A disadvantage of the OSA notation is that much may be lost when a minus sign is overlooked or omitted. Quantitative definitions of L, j, and g are given in [Ref. 4.4, p. 165].

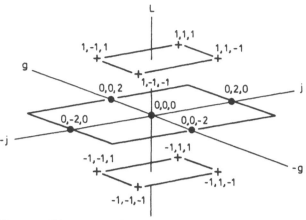

Fig. 9.5. The 13 central points of the basic set in OSA color space (corresponding to the corner points and central point of the cubo-octahedron in Fig. 9.1) and their OSA color notations (L, j, g)

The notations for the central 13 points in OSA color space are given in Fig. 9.5. The central point O is indicated by $(0, 0, 0)$. Also we see that, in planes at odd-numbered lightness levels $L = 1$ and $L = -1$, points are located at odd-numbered coordinates: $j = 1, g = 1; j = 1, g = -1; j = -1, g = 1;$ and $j = -1, g = -1$. This is a general characteristic of the basic set. Furthermore, as a consequence of the regular alternation of the locations of grids mentioned above, at even-numbered levels $(L = 2, 4; -2, -4, -6)$ and at level $L = 0$, the coordinates of the locations of points are even numbers, for example, $j = 2, g = 2;$ and $j = 0, g = -4$. These situations are summarized by the grids in Figs. 9.6 and 7 which show points (o) at the even-numbered lightness levels and points (+) at the odd-numbered lightness levels, respectively. Referring to Figs. 9.1 and 2, we see that point $J(0, 2, 0)$ is represented in Fig. 9.6 and points $A(-1, 1, 1)$ and $N(1, 1, -1)$ are shown in Fig. 9.7.

The neutral-gray samples are located at $j = 0, g = 0$ at lightness levels $L = -6, -4, -2, 0, 2, 4$. The grays become lighter in uniform steps with lightness L increasing from -6 to 4. At the central lightness level $L = 0$, the gray sample is a medium gray that has a luminance factor $Y_{10} = 0.30$ (CIE ILL D$_{65}$), which corresponds approximately to Munsell Value $V = 6$ (CIE ILL C). Although the grays are given in perceptually equal steps, they are not nearest neighbors, because they are represented on only alternate (even-numbered) lightness levels (Table 9.1). Colors that produce equal L but different hue or saturation do not usually have equal Y_{10} or Munsell Value [8.29,32,33; 9.3].

All the colors of the basic set of samples, which are limited to pigments of adequately high permanency [5.11], are represented at lightness levels from $L = 5$ (Munsell Value 8.7 on the neutral axis, $j = 0, g = 0$) down

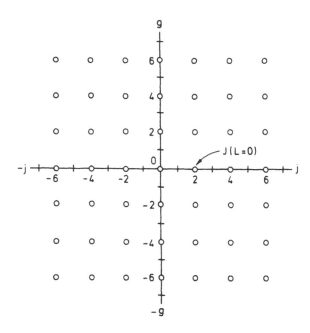

Fig. 9.6. Square-grid arrangement of even-numbered coordinates (points o) in even-numbered horizontal planes ($L = -6, -4, -2, 0, 2, 4$)

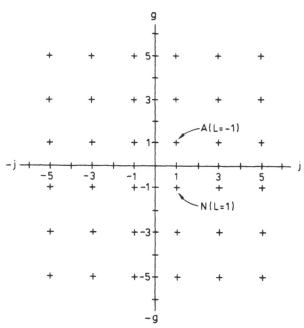

Fig. 9.7. Square-grid arrangement of odd-numbered coordinates (points +) in odd-numbered horizontal planes ($L = -7, -5, -3, -1, 1, 3, 5$)

Table 9.1. The OSA-UCS neutral-gray samples [8.29,30]

L (nominal)	-6	-4	-2	0	2	4
L (actual)	-6.1	-3.9	-2.0	0.1	2.1	4.1
V (actual)	2.72	3.96	5.04	6.01	7.03	8.15
Y_{10} (actual)	0.054	0.117	0.201	0.302	0.435	0.618
ISCC-NBS color (Sect. 10.1)	Dark gray	Dark gray	Medium gray	Medium gray	Light gray	Light gray

L : OSA lightness (CIE ILL D_{65})
V : Munsell Value (CIE ILL C)
Y_{10}: luminance factor (CIE ILL D_{65})

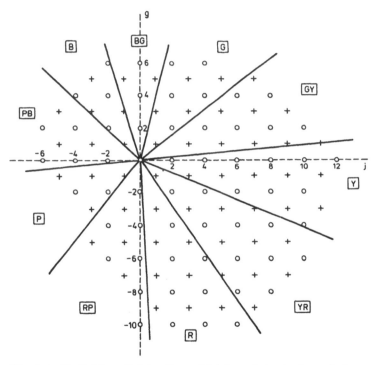

Fig. 9.8. Radial hue-division lines. Division of the horizontal cleavage planes into 10 major Munsell Hue sectors. The points given by coordinates (j, g) indicate the nominal location of the colors of the OSA samples on horizontal planes. ○: Even-numbered coordinates and planes; +: odd-numbered coordinates and planes [Notes 9.2,3]

to $L = -7$ (Munsell Value 2.3 on the neutral axis), which are the lightness limits of the set. Figure 9.8 shows sectors that give a good idea how Munsell Hue ranges (R, YR, Y, GY, etc.) are distributed on a horizontal plane [Notes 9.2,3]. All color-sample points, for both odd-numbered (+) and even-numbered (○) lightness levels, are included at nominal locations on the diagram.

9.3 Uniform Color Arrays on the Series of Horizontal Planes Through OSA Color Space

If we wish to make a color chart that shows a two-dimensional array of perceptually equally spaced colors, we should consider the variety of uniform arrays that can be found on the various cleavage planes that pass through the color-sample points in OSA color space. There are seven different series of parallel cleavage planes, a total of about 80 planes, each of which produces a uniform array.

One series consists of 13 equally spaced *horizontal cleavage planes,* which pass through all the points that represent color samples. (Spaces between planes are visible in Plate XII.) This series of planes provides the basis for the systematic location of colors in color space and for a color notation for their identification. In the technical literature on the OSA system, the 13 horizontal planes are commonly designated by their corresponding lightness levels: $L = -7$, $L = -6$, ..., $L = 0$, ..., $L = 4$, $L = 5$. To conform more closely with the notation adopted in this book for the vertical and oblique cleavage planes, the horizontal planes are designated here by: $L(-7)$, $L(-6)$, ..., $L(0)$, ..., $L(4)$, $L(5)$ [Note 9.1].

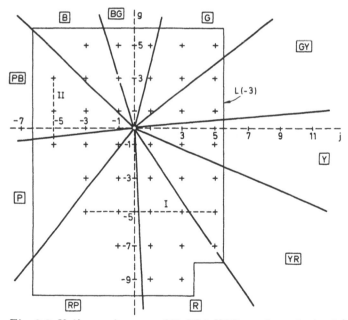

Fig. 9.9. Uniform color array of 39 OSA-UCS samples on horizontal cleavage plane $L(-3)$ (at lightness level $L = -3$), see Plate XIII. The radial hue-division lines indicate the 10 major Munsell Hue regions (Fig. 9.8). This diagram is part of that in Fig. 9.11; it illustrates the use of Figs. 9.10 and 11. Two uniform color scales are indicated: *I,* $(-3, -3, -5)$ $(-3, 5, -5)$ and *II,* $(-3, -5, -1)$ $(-3, -5, 3)$

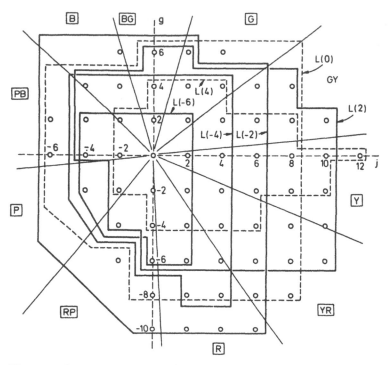

Fig. 9.10. Uniform color arrays on the even-numbered horizontal cleavage planes $L(-6)$, $L(-4)$, $L(-2)$, $L(0)$, $L(2)$, $L(4)$. The radial hue-division lines indicate the 10 major Munsell Hue regions (Fig. 9.8)

The uniform color arrays of the horizontal planes are of particular interest because each displays color samples of the same lightness. One such uniform color array is shown in Plate XIII. The square-cut color samples have been mounted on equally spaced center points on a square grid to display equal distances to nearest neighbors. The 39 colors are described by their lightness level $L = -3$ and the coordinates (j, g) of the diagram shown in Fig. 9.9. Also shown in Fig. 9.9 are the radial hue-division lines presented in Fig. 9.8.

In [Ref. 9.7, Tables I–XIII], a nominal indication of the color samples that are included in each horizontal plane is provided in tabular form. This information is presented here in two diagrams: Fig. 9.10, which shows points (∘) that represent all colors in the arrays presented in the even-numbered horizontal planes [$L(-6)$ to $L(4)$] and Fig. 9.11, which shows points (+) for all colors in arrays in odd-numbered horizontal planes [$L(-7)$ to $L(5)$].

To illustrate the use of Figs. 9.10 and 11, let us find which OSA color samples are available at lightness level $L = -3$ (a rediscovery of the 39 samples discussed above). We consult Fig. 9.11 because it applies to odd-numbered horizontal planes. The desired array is located by noting its sur-

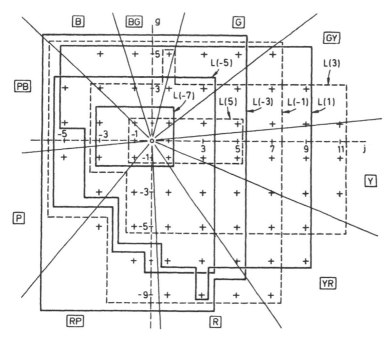

Fig. 9.11. Uniform color arrays on the odd-numbered horizontal cleavage planes $L(-7)$, $L(-5)$, $L(-3)$, $L(-1)$, $L(1)$, $L(3)$, $L(5)$. The radial hue-division lines indicate the 10 major Munsell Hue regions (Fig. 9.8)

rounding boundary line labelled $L(-3)$. This array and boundary line are reproduced in Fig. 9.9. Each color in the array has a different OSA chromaticness designation (j, g). For example, for one color we find on Fig. 9.9 that $j = 5$ and $g = 5$. Because the lightness level is $L = -3$, the OSA notation (L, j, g) for the color is $(-3, 5, 5)$. The radial division lines show that the hue is a green. The relatively large distance of the point from the central zero point indicates a relatively high perceived saturation.

9.4 Uniform Color Arrays on the Two Series of Parallel Vertical Planes Through OSA Color Space

In addition to the set of 13 horizontal planes discussed in the previous section, there are two sets of *parallel vertical cleavage planes* that pass through all the points occupied by the colors of the samples in OSA color space. (And, as we shall see in Sect. 9.6, there are also four sets of parallel oblique planes that pass through all the points.) The planes of one set of parallel vertical planes are perpendicular to those of the other set, and, of course, all are perpendicular to the horizontal planes.

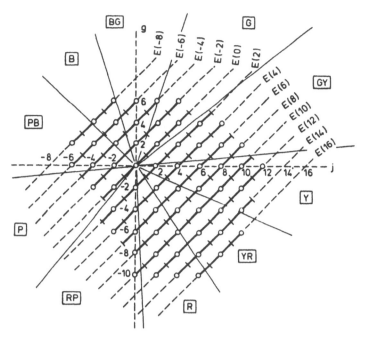

Fig. 9.12. Notation for the vertical cleavage planes (series *E*). Each is identified by its intersection with a horizontal plane and with the *j* axis. A heavy line indicates the part of a vertical plane occupied by a uniform color array [even-numbered (○) and odd-numbered (+) coordinates (j, g)]. The radial hue-division lines indicate the 10 major Munsell Hue regions (Fig. 9.8)

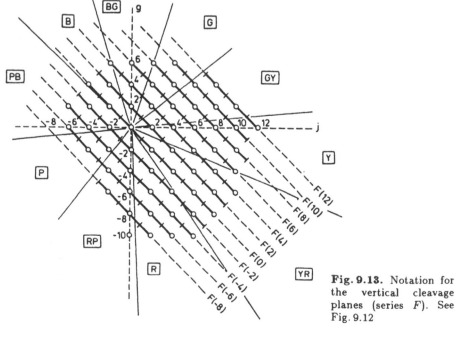

Fig. 9.13. Notation for the vertical cleavage planes (series *F*). See Fig. 9.12

To visualize the vertical planes, it is helpful to refer to the cubo-octahedron in Fig. 9.1. One vertical plane passes through points N, R, D, and J. In Figs. 9.1,2, it is obvious that the plane is perpendicular to the horizontal planes. A parallel vertical plane passes through E, M, C, A, and O, and another vertical plane passes through F, B, K, and G. These three vertical planes belong to a series of 13 parallel vertical planes that pass through all the points that represent color samples in OSA color space. In this book, this series of vertical planes is called the E series, because the plane that passes through the neutral axis also passes through point E (Fig. 9.1).

Figure 9.12 shows a horizontal plane with slanted lines drawn on it to represent traces or cuts made by the 13 vertical planes of the E series that pass through it. The individual planes of the series are designated by the points where they cut the j axis. Hence the vertical plane of the E series that cuts the j axis at $j = -8$ is assigned the notation $E(-8)$; the one that cuts the axis at $j = 0$ is represented by $E(0)$, and so on. In technical literature on the OSA system, the vertical planes are represented by mathematical equations that define the straight lines made by the cuts in a horizontal plane [Note 9.1].

Let us now consider the planes of the second series of parallel vertical cleavage planes, series F, which, as noted above, are perpendicular to those forming series E. In Fig. 9.1, one plane of the F series passes through points M, B, C, and R. Another passes through F, N, D, K, and O, and another through E, G, A, and J. In this case, the series is identified by F because the vertical plane that passes through the neutral axis also passes through point F. There are 11 vertical planes in the F series; their traces or cuts on a horizontal plane are indicated by the slanted lines in Fig. 9.13. Here again the designation of a vertical plane is made by indicating where it cuts the j axis on a horizontal plane. Hence, plane $F(12)$ cuts the j axis at $j = 12$.

The 13 color arrays available in the E series are represented on a chart (Fig. 9.14), where, in a manner similar to the presentation used in [9.7], points are plotted on a grid that shows g versus L for each of the planes of the E series. At the top of the chart, the corresponding values of j are tabulated for each vertical plane, within the range for which color samples are available. The representations of the arrays on the 13 vertical planes are shown superposed one on another. The boundaries of the arrays are labelled with the designations of the planes to which they belong. Thus, for example, the boundary of the array in vertical plane $E(-8)$ is labeled $E(-8)$ (Fig. 9.14); the portion of the diagram that shows only this array is reproduced in Fig. 9.15. The symbols for even-numbered (∘) and odd-numbered (+) lightness levels appear in the arrays. Figure 9.16 presents the 11 arrays available in the F series of parallel vertical planes.

The positions of the heavy, slanted lines with respect to the j and g axes on a horizontal plane (Figs. 9.12, 13) suggest the OSA chromaticness range (perceived saturation and hue) within which colors are available in

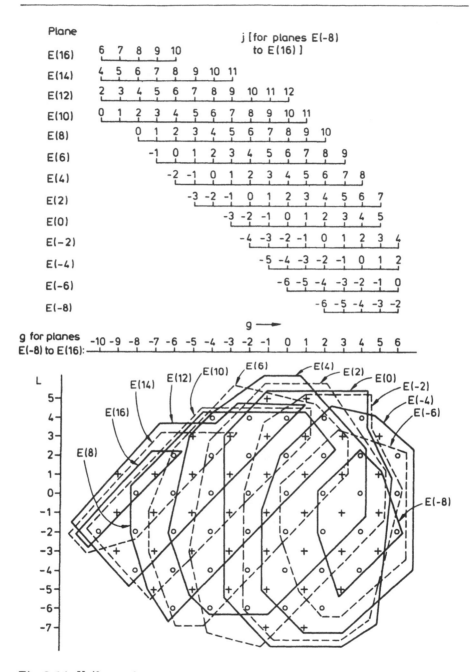

Fig. 9.14. Uniform color arrays on the 13 vertical cleavage planes of series E

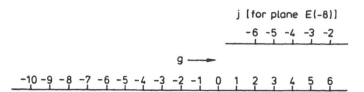

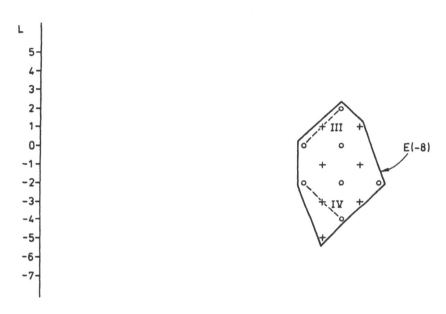

Fig. 9.15. Uniform color array on the vertical cleavage plane $E(-8)$. Part of Fig. 9.14 to illustrate the use of Figs. 9.14 and 16. Two uniform color scales are indicated: *III*, $(0, -6, 2)$ $(2, -4, 4)$, and *IV*, $(-4, -4, 4)$ $(-2, -6, 2)$

the collection of samples. The radial hue-boundary lines aid in visualizing the hues in a range. The arrays that appear in vertical planes $E(0)$ and $F(0)$ are of particular interest because they are the only ones that include the entire series of neutral grays. And, in these two arrays, near-complementary (additive) hues are represented: green and purple in plane $E(0)$; blue and orange in $F(0)$ (Sect. 9.8).

In a vertical plane, each point that represents a color has four nearest-neighbor points in the plane (Fig. 9.15). Thus the points can be plotted on a square grid much like the points on a horizontal plane. A unit square grid for the E series of vertical planes is suggested by the square NRDJ in Fig. 9.1, and one for the F series by the square MBCR. But note that the square grids are tilted (just as the unit squares are). They show the nearest neighbors placed along slanted lines. The points shown in Fig. 9.15

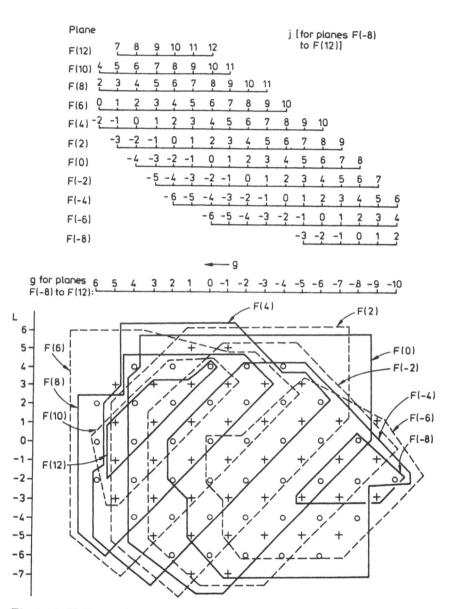

Fig. 9.16. Uniform color arrays on the 11 vertical cleavage planes of series F

could serve as the center points of square-cut color samples mounted on a chart to display the array of vertical plane $E(-8)$. The dashed lines shown emphasize the tilted arrangement of the nearest-neighbor color samples. Of course, for display purposes on a chart, the array of square color samples is generally turned through 45° to show horizontal rows and vertical columns of color samples without disturbance to the relationships between the colors.

9.5 OSA Uniform Color Scales (OSA-UCS)

A *uniform color array* is a two-dimensional arrangement of color samples, each differing from its nearest neighbors by a perceptually equal amount. In principle, a color array could consist of a minimum of three colors, but in the OSA set of samples, on most planes, the arrays consist of more samples, often a network of *uniform color scales*. Each uniform color scale is a linear (straight-line) sequence of three or more colors (in OSA space) that have equal perceived color differences. The 13 arrays provided on the series of horizontal cleavage planes and the 24 arrays on the two series of vertical cleavage planes contain in their networks all the uniform color scales that can be formed from the OSA basic set. For this reason, these 37 arrays, represented in Figs. 9.10 and 11 for the horizontal planes and in Figs. 9.14 and 16 for the two series of parallel vertical planes, serve as practical sources for locating and specifying *all the uniform color scales*.

In a horizontal plane, the colors in the uniform color scales and the uniform color arrays are characterized by constant lightness L. Thirteen uniform color scales can be found, for example, in the uniform color array displayed in Plate XIII. These can be identified in Fig. 9.11 in the array on horizontal plane $L(-3)$ or in Fig. 9.9, where the array has been isolated. The uniform scales can be revealed by drawing straight lines through nearest-neighbor points (a minimum of three) parallel to the j and g axes. In Fig. 9.9, dashed lines are shown drawn through two such sequences of points, to illustrate two uniform scales (I and II) of the array in Plate XIII.

Any color scale in OSA color space can be identified by the notation (L, j, g) for the color at each end of the scale. This method of designating scales has been adopted here [Note 9.1]. Thus, color scale I (Fig. 9.9) is identified by $(-3, -3, -5)$ $(-3, 5, -5)$. For color scales in horizontal planes, the end point with the lower value of j or g (whichever is not constant) is listed first. [Said differently, the end point that is seen on the diagram to be either lower than or farther to the left of the other end point is listed first.] This identification for I reveals that there are 5 colors in the scale, because the one parameter that changes, j, has 5 different values: -3, -1, 1, 3, 5. In scale II, identified by $(-3, -5, -1)$ $(-3, -5, 3)$, there are three colors because the parameter that varies, g, has three different values: -1, 1, 3.

The two values of L in the designation of a color scale are identical (for example -3 in scale I, also in scale II) if the color scale is of uniform lightness [the color scale is located on a horizontal cleavage plane, here $L(-3)$]. The procedure for finding a horizontal array containing a particular color scale involves the use of Figs. 9.10 and 11, as described above.

Each uniform color scale in a vertical array (in a vertical cleavage plane) consists of a sequence of colors that progresses in a stepwise manner from one lightness level to the next. The color array present in vertical plane

$E(-8)$ in Fig. 9.14 and shown in isolation in Fig. 9.15 contains seven uniform color scales. They may be found by connecting sequences of three or more nearest-neighbor points by straight lines. It must be remembered that these straight lines are slanted; the nearest-neighbor points in a vertical cleavage plane fall on the intersections of a tilted square grid.

Before considering how a uniform color scale in a vertical cleavage plane can be designated, let us first see how a point is specified. The topmost point in the array in vertical plane $E(-8)$ will serve as an example. Included in the information transferred from Fig. 9.14 to 15 is the numerical scale for j which applies specifically to cleavage plane $E(-8)$. The point that is considered represents the color that has lightness $L = 2$, from the scale at the left, and $j = -4$ and $g = 4$, from the two scales above. Hence the designation of the color is $(2, -4, 4)$.

Two uniform color scales (III, IV) in plane $E(-8)$ are indicated by dashed lines (Fig. 9.15). Here again, as in the case of uniform color scales in horizontal planes, a color scale is designated by a notation that specifies the color at its two ends. Thus, for scales III and IV we have $(0, -6, 2)$ $(2, -4, 4)$ and $(-4, -4, 4)$ $(-2, -6, 2)$, respectively. Note that, in each case, the notation lists first the end point that has the lower value of L – that is, the end point that is seen on the diagram to be lower. The number of colors in a uniform color scale *in a vertical plane* is indicated by the range of values through which L is varied. In scales III and IV, L has three values (0, 1, and 2; -4, -3, and -2, respectively); hence, each scale has three colors.

If we wish to select uniform color scales (in vertical planes) that have specific hues, we should first consult Figs. 9.12 and 13 to find uniform arrays that contain them. These two diagrams contain the radial hue-boundary lines, which indicate the Munsell Hue zones, traversed by uniform color arrays (represented by heavy lines). The small circles (o) and crosses (+) along a heavy line follow the style adopted for Figs. 9.10 and 11 to differentiate between sample points located on even-numbered horizontal planes and those located on odd-numbered horizontal planes (Figs. 9.6, 7).

9.6 Uniform Color Arrays on the Four Series of Parallel Oblique Planes Through OSA Color Space

The one series of horizontal cleavage planes and the two series of parallel vertical cleavage planes are important because they present not only interesting uniform color arrays but also three series of different uniform color scales which are the only scales that exist in OSA color space. The four series of *parallel oblique cleavage planes* present no additional uniform color scales, but they do show 41 different color arrays composed of uniform color scales.

To aid us in visualizing the oblique planes, let us refer once again to the central cubo-octahedron in OSA color space (Fig. 9.1). The oblique planes are tilted at 55° with respect to the horizontal and vertical planes. One oblique plane rises toward the central axis through point R and points M and N. A second plane parallel to the first rises through points C and D, points B, O, and J, and points F and E. A third parallel plane rises through points K and A and through point G. These three planes are in one series of parallel oblique planes. Because this series of planes rises from the portion of OSA color space characterized generally by redness and one plane rises through point R toward the central axis, this series is represented in this book as the R series [Note 9.1]. Similarly, the J series, which rises from the portion characterized by yellowness, includes one plane that rises through point J and points E and N; a second plane of the series rises through points A and D, points G, O, and R, and points F and M, etc. Series G, which rises from the zone of greenness, and series B, which rises from the zone of blueness, are illustrated in the same way.

We can see in Fig. 9.1 that all the oblique planes of the J and B series cut the j axis but not the g axis. Similarly, all oblique planes of the R and G series cut the g axis but not the j axis. The oblique planes are designated by their series-letter symbol and by the lightness level L at which they intersect the central L axis. Thus, $J(2)$ designates a plane of the J series that passes through the point on the L axis identified by $L = 2$. Plane $B(2)$ is one of the B series; it passes through the same point $(L = 2)$ on the central axis, and $R(0)$ is a plane of the R series that crosses the central axis (at the central point) at $L = 0$.

The "edge-on" profiles of the J and B parallel oblique cleavage planes are shown in Figs. 9.17 and 18. The points at which they cross the central axis are clearly evident. Two similar sets of profiles for the G and R parallel oblique cleavage planes are presented in Figs. 9.19 and 20, which again show the crossing points on the central axis. The heavily drawn portions of the slanted lines represent "edge-on" views of uniform color arrays that contain two or more uniform color scales. [Planes $J(16)$ and $R(10)$ each contain only one color sample. Plane $J(14)$ contains five color samples, three of which form one horizontal uniform color scale.] The locations of the heavy lines with respect to the various regions characterized by yellowness, greenness, blueness, and redness can aid in the selection of uniform color arrays (Fig. 9.8).

To get an idea of how to produce a uniform color array in an oblique plane, let us consider an array of primarily bluish character, for example the one found in oblique plane $B(4)$ (Fig. 9.18). An array in an oblique plane can be built up from a series of uniform color scales (sometimes with the addition of single colors or color pairs at the boundaries of the array) from either the horizontal series of planes or from the vertical series. It is easier to employ the horizontal series in this task. Hence, we shall need

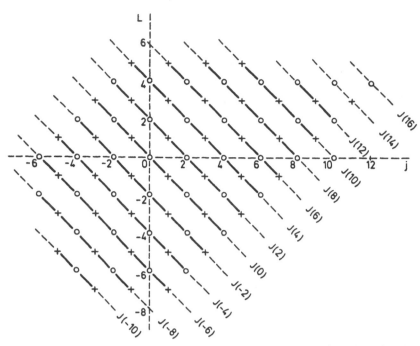

Fig. 9.17. Notation for the 14 oblique cleavage planes of series J. Each is identified by its "edge-on" profile and its intersection with the L axis. A heavy line indicates the part of an oblique plane occupied by a uniform color array [even-numbered (○) and odd-numbered (+) coordinates (j, g)] [Note 9.4, p. 246]

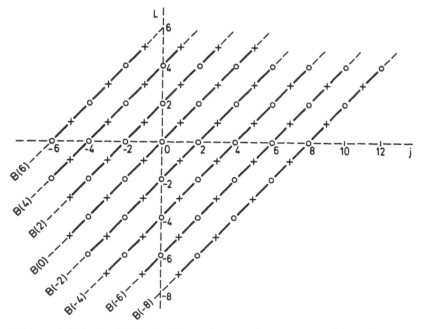

Fig. 9.18. Notation for the 8 oblique cleavage planes of series B. See Fig. 9.17

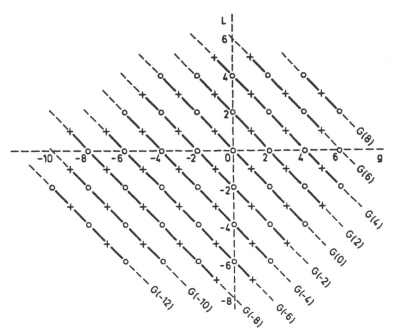

Fig. 9.19. Notation for the 11 oblique cleavage planes of series G. Each is identified by its "edge-on" profile and its intersection with the L axis. A heavy line indicates the part of an oblique plane occupied by a uniform color array [even-numbered (o) and odd-numbered (+) coordinates (j, g)] [Note 9.4, p. 246]

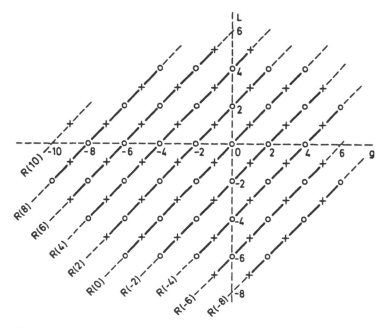

Fig. 9.20. Notation for the 10 oblique cleavage planes of series R. See Fig. 9.19

both Figs. 9.10 and 11 and, as a guide, Fig. 9.18. The procedure resembles the building of a roof: The slanted line labeled $B(4)$ in Fig. 9.18 may be taken as the "edge-on" profile of the rafters. Planks need to be placed horizontally across the rafters at points designated by (○) and (+). In the color array, the planks correspond to uniform color scales (and sometimes single colors and color pairs) obtained from Fig. 9.10 (○) and Fig. 9.11 (+). The development of the array in this way (from its bottom row to its top row) may be followed in Fig. 9.21.

The bottom row of the array [see the line labeled $B(4)$, which represents the oblique plane in which the array is found (Fig. 9.18)] contains colors of lightness $L = -2$, an unchanging j ($j = -6$), and a sequence of values of g. Figure 9.10 shows, however, that in horizontal plane $L(-2)$ and at $j = -6$, there is only a color pair, for which $g = 0$ and $g = 2$. These two points are shown plotted at the bottom of the array (Fig. 9.21). The next step is to consider points for which $L = -1$ and $j = -5$ (Fig. 9.18). In Fig. 9.11, we

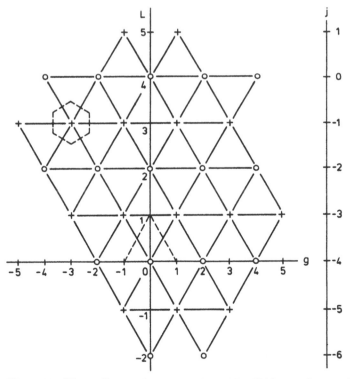

Fig. 9.21. The uniform color array of 31 OSA-UCS samples on oblique cleavage plane $B(4)$. Seventeen uniform color scales are illustrated by the slanted and horizontal lines that connect points. The uniform color scales can be designated by the notation (L, j, g) of their end points (Sect. 9.5). Values of j are indicated at the right [they were obtained from the line for oblique cleavage plane $B(4)$ in Fig. 9.18]

can see that at $L = -1$ a uniform color scale is formed by the three points $g = -1$, $g = 1$, and $g = 3$ at $j = -5$. This scale constitutes the second row at the bottom of the array. The procedure is continued until the limit of the available colors is reached – that is, in this case, two samples at $L = 5$ and $j = 1$ (Fig. 9.11).

The uniform color array of oblique cleavage plane $B(4)$ is shown in Fig. 9.21 by points connected in a manner to reveal six horizontal and eleven slanted uniform color scales. This array shows that in the oblique cleavage plane each point is surrounded by six nearest neighbors, all of which are equally far from it, which corresponds to equal perceptual differences. The six points describe a hexagon, which is composed of six equal-sided triangles. If, when a chart of such an array is made, the samples are cut in a hexagonal shape (see the hexagon formed by dashed lines) and mounted at the designated points, the samples can be oriented to show equal spacing between nearest neighbors. The grid on which the samples are mounted is not a square grid, however; it is a rectangular grid. Such a grid was made in drawing Fig. 9.21. The construction of the grid was begun by employing the line between $g = -1$ and $g = 1$ on the g axis as the base of an equal-sided triangle. After constructing the triangle, two sides of which are shown as dashed lines, its top point (apex) established the point $L = 1$ on the central axis. The lightness L scale can then be measured along the central axis, and the lines of the rectangular grid can then be drawn in. This procedure for establishing a uniform color array in a plane in the B series of oblique planes can be applied directly in establishing one in a plane in the J series. The procedure is much the same for the G and R series. The minor difference is that the planks used in building the roof represent uniform color scales, each at an unchanging value of g and a sequence of values of j.

As stated above, all the uniform color scales of the OSA-UCS system appear without repetition on the horizontal and vertical cleavage planes. Each scale reappears twice in different combinations with other scales on the oblique cleavage planes. This can be understood if we examine the intersections of one plane from the series of horizontal planes and of one plane from each of the two series of vertical planes with one plane from each of the four sets of oblique planes. Each intersection occurs along a straight line. Because the straight line is part of the two planes that form it and because the straight line corresponds to the location of a uniform color scale, the color scale must be present in both planes.

The systematic reappearance of horizontal color scales in opposing pairs of oblique planes (planes J and B, and planes G and R) that rise through the horizontal planes is easily followed; the idea was employed above, following the example of placing planks on rafters. Precisely the same planks used on the rafters designated by J are used on rafters designated by B. Specifically, each color scale (plank) of constant j appears both on a J and on a B plane; each scale (plank) of constant g appears both on a G and on a R plane.

To illustrate the distribution of slanted color scales in the oblique planes, hold this book vertically and opened at 90°. The page on the left represents vertical plane $F(0)$; that on the right, $E(0)$ (Figs. 9.12,13). A straight line drawn upwards (45°) on the left-hand page and another straight line inclined (45°) upwards on the right-hand page meet at a point on the vertical line where the pages join. The vertical line represents the central axis. The two slanted lines drawn on the pages indicate where an oblique plane [say plane $J(0)$, Fig. 9.17] rising from the yellowness zone facing the opened pages cuts through the two vertical planes and where it cuts the central axis [at $L = 0$]. Each line formed by the intersection of cleavage planes in OSA color space represents the location of an available (or unavailable) uniform color scale. Clearly, the color scale in the vertical plane on the left, $F(0)$, is found also in oblique plane $J(0)$. It is found as well in adjoining plane $R(0)$, because, rising from the zone of redness at the left, plane $R(0)$ intersects vertical plane $F(0)$ along the same line (but on the other side of the page at the left). Another color scale, juxtaposed below the first one on vertical plane $F(0)$, would appear on lower oblique planes of the same two series – namely, planes $J(-2)$ and $R(-2)$.

Now let us consider a color scale that is perpendicular to the first color scale, on vertical plane $F(0)$ (on the page on the left), for example, a scale described by a line that moves downwards at 45° to the point $L = 0$ on the central axis, where the first pair of drawn lines converge. Following the same reasoning, we see that a perpendicular line (and a corresponding color scale) on $F(0)$ reappears on both oblique planes $B(0)$ and $G(0)$. They rise from behind the book, in the blueness and greenness zones. A second perpendicular line (and uniform color scale) juxtaposed below the first one on $F(0)$ appears on planes $B(-2)$ and $G(-2)$.

9.7 The Pastel Color Samples

The pastel colors at the center of the basic set are rather few in number. In order to increase their number and thereby to provide a more useful representation, it was decided to introduce intermediate samples, bisecting color differences in the lightness range $L = -2$ to $L = 2$ and within the ranges of chromaticness given by $g = -3$ to $g = 3$ and $j = -3$ to $j = 3$ [Refs. 4.4, p. 174; 8.29,30,33]. The resulting half-step set, or *pastel set*, utilizes 51 colors from the basic set and 134 intermediate colors. The horizontal unit square grids at the center of the pastel set are shown at three lightness levels ($L = -0.5, 0, 0.5$) in Fig. 9.22.

The values of j and g that represent colors at all lightness levels of the pastel set may be found in Figs. 9.23 and 24. Radial hue-boundary lines are shown, which divide the horizontal planes into 10 Munsell Hue zones (Fig. 9.8). Uniform color arrays and uniform color scales in the horizontal

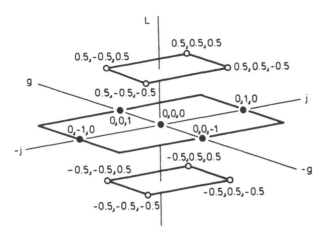

Fig. 9.22. The 13 central points of the pastel set in OSA color space (analogous to the corner points and central point of the cubo-octahedron in Fig. 9.1) and their OSA color notations (L, j, g)

cleavage planes can be selected from these diagrams in much the same way as described in Sects. 9.3 and 5. In Fig. 9.23, the dashed square that encloses all the points comprises arrays in planes $L(0.5)$ and $L(-0.5)$; the inner square comprises arrays in planes $L(1.5)$ and $L(-1.5)$.

All vertical cleavage planes through the pastel set are represented in Figs. 9.25 and 26, which are related to those shown in Figs. 9.12 and 13. Because all of the points that represent the pastel colors occupy a symmetrical volume in color space, as arrays on vertical planes of series E' and F' they can conveniently be presented on one diagram (Fig. 9.27). There are two separate scales for g; one is for the E' series and the other for the F' series. But for any given value of g, at any level L, the value of j varies from one plane to another in a series. The ranges of values of j are given at the top of Fig. 9.27. The range of values of j for plane $E'(-3)$, for example, is positioned over the points of an array shown below and labeled $E'(-3)$.

To specify a color given by a point in the array, we note its lightness level L in the column at the left and its values for j and g in the appropriate rows above. Thus, for the top point in the tilted square array $E'(-3)$, we note $L = 1.5$, $j = -1.5$, $g = 1.5$. Hence the specification for the color is $(1.5, -1.5, 1.5)$. The uniform color scales on the vertical planes are identified in the manner described in Sect. 9.5.

The four series of parallel oblique cleavage planes that pass through the pastel set are related to those in the basic set (Sect. 9.6). The procedure for building an array in an oblique plane is the same as that described for the basic set. Here Figs. 9.23 and 24 are required. Figures 9.28 and 29 (which are related to Figs. 9.17–20) are presented to serve as an aid.

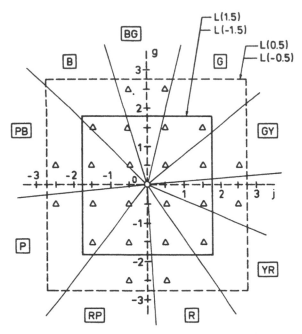

Fig. 9.23. Four uniform color arrays on the fraction-numbered horizontal cleavage planes of the pastel set $L(-1.5)$, $L(-0.5)$, $L(0.5)$, $L(1.5)$. The radial hue-division lines indicate the 10 major Munsell Hue regions (Fig. 9.8)

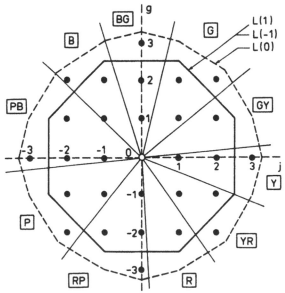

Fig. 9.24. Three uniform color arrays on horizontal cleavage planes of the pastel set $L(-1)$, $L(0)$, $L(1)$. The radial hue-division lines indicate the 10 major Munsell Hue regions (Fig. 9.8)

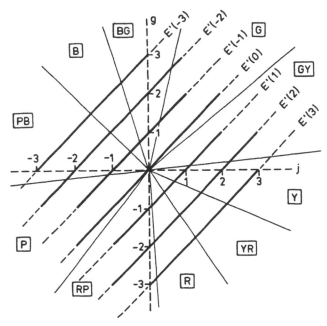

Fig. 9.25. Notation for the 7 vertical cleavage planes of series E', pastel set. Each is identified by its intersection with a horizontal plane and with the j axis. A heavy line indicates the part of a vertical plane that is occupied by a uniform color array. The radial hue-division lines indicate the 10 major Munsell Hue regions (Fig. 9.8)

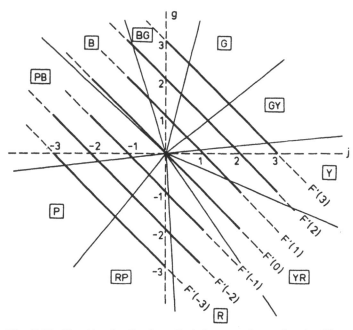

Fig. 9.26. Notation for the 7 vertical cleavage planes of series F', pastel set. See Fig. 9.25

j for planes
E'(-3) to E'(3) and
F'(-3) to F'(3)

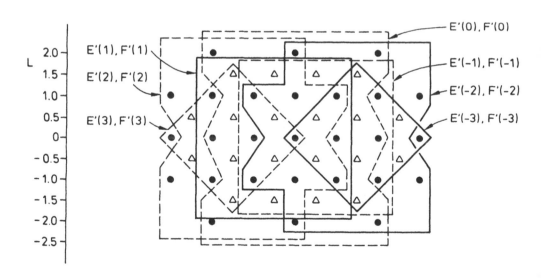

	j						
E'(3),F'(3)	0 0.5 1 1.5 2 2.5 3
E'(2),F'(2)	-1 -0.5 0 0.5 1 1.5 2 2.5 3
E'(1),F'(1)	-1 -0.5 0 0.5 1 1.5 2
E'(0),F'(0)	-2 -1.5 -1 -0.5 0 0.5 1 1.5 2
E'(-1),F'(-1)	-2 -1.5 -1 -0.5 0 0.5 1
E'(-2),F'(-2)	-3 -2.5 -2 -1.5 -1 -0.5 0 0.5 1
E'(-3),F'(-3)	-3 -2.5 -2 -1.5 -1 -0.5 0

g ⟶

g for planes
E'(-3) to E'(3): -3 -2.5 -2 -1.5 -1 -0.5 0 0.5 1 1.5 2 2.5 3

⟵ g

g for planes
F'(-3) to F'(3): 3 2.5 2 1.5 1 0.5 0 -0.5 -1 -1.5 -2 -2.5 -3

Fig. 9.27. Uniform color arrays in all the vertical cleavage planes of the pastel set (series E' and F')

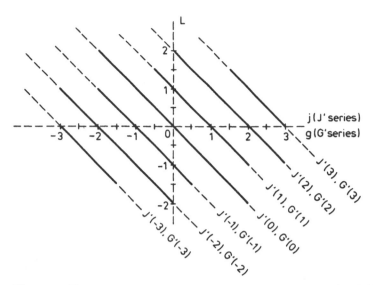

Fig. 9.28. Notation for the 7 oblique cleavage planes of series J' and G', pastel set. Each is identified by its "edge-on" profile and its intersection with the L axis. A heavy line indicates the part of an oblique plane occupied by a uniform color array. [Note 9.4, p.246]

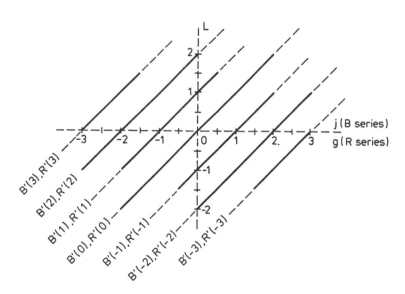

Fig. 9.29. Notation for the 7 oblique cleavage planes of series B' and R', pastel set. See Fig. 9.28

169

9.8 Comments About the OSA Samples

In the above sections, I have attempted to present clearly the arrangement of colors in OSA color space. The approach employed depended upon the cubo-octahedron as a key to the arrangement and a description of the cleavage planes and, thereby, of the color scales themselves.

If the selection of colors were made on the bases of a lattice of cubes (instead of interlocking cubo-octahedra), as in CIELUV and CIELAB color spaces (Sect. 8.2), the array of points in color space would be comparatively simple to visualize. But, in such an array, each color would be surrounded by only six equally spaced nearest neighbors, whereas in the OSA-UCS system each is surrounded by twelve, which is the maximum possible in three-dimensional space.

In the OSA-UCS system, seven different series of parallel cleavage planes are available for displaying color scales. In a cubical array there are only three series. In the Munsell color system there is just one series of parallel cleavage planes: a horizontal series perpendicular to the Value axis.

In the Munsell color system, a horizontal cleavage plane reveals a Munsell Hue-Chroma diagram (Fig. 8.10), which displays at one Value level equally spaced concentric color circles and equally spaced radial lines whose intersections are intended to represent samples whose perceived colors possess one Chroma and equally spaced Hues in each circle, and one Hue and equally spaced Chromas along each radial line [Note 8.1]. In the OSA-UCS system such representation by color samples in accordance with similar parameters is much more limited. In a horizontal plane in the basic set, circles centered on point $(j = 0, g = 0)$ can be drawn through points that represent colors of approximately constant saturation, but, although the hues are essentially equidistant, the colors are not nearest neighbors (except in the case of the four colors of lowest saturation in the odd-numbered horizontal planes of the basic set or in the fraction-numbered horizontal planes of the pastel set). In each of the oblique series of cleavage planes, circles can also be drawn centered on point $(j = 0, g = 0)$. (All the oblique series intersect the central axis at even-numbered values of lightness L.) In such a case, the smallest circle in each plane passes through the points for the six nearest-neighbor colors [for example, points F, M, R, D, A, and G in oblique plane $J(0)$ in Fig. 9.1] whose hues are equally spaced, perceived saturation is constant, and lightness is varied.

Two color scales of complementary (additive) hues and unchanging lightness can be found at all even-numbered lightness levels in OSA color space. Two such scales are indicated by points (at $L = 0$) on the CIE 1964 (x_{10}, y_{10}) chromaticity diagram in Fig. 9.30 [8.29]. The closeness of these points to a linear alignment shows that the hues are essentially complementary. Uniform color scales of varying lightness that include neutral gray can be found in the uniform color arrays in two vertical cleavage planes in

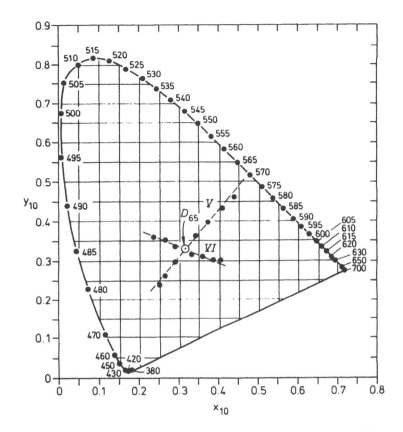

Fig. 9.30. Two OSA uniform color scales, in the horizontal plane $L(0)$, containing near-complementary (additive) hues. Color scales V, $(0, -6, 0)$ $(0, 8, 0)$, and VI, $(0, 0, -8)$ $(0, 0, 6)$. CIE 1964 (x_{10}, y_{10}) chromaticity diagram

OSA color space (Sect. 9.4). From these color arrays, complementary pairs of green and purple, and of orange and blue hues can be selected. An estimation of the identity of these hues can be made from the positions of vertical planes $E(0)$, $E'(0)$, $F(0)$, $F'(0)$ relative to the radial hue-boundary lines (Figs. 9.12,13,25,26).

In the basic set there are 414 uniform color scales and 78 uniform color arrays (containing at least one uniform color scale). Among the color scales, 143 are of constant lightness. Twelve uniform color scales of constant lightness and 24 scales of varying lightness include a neutral gray. Indeed, a marked advantage of the OSA-UCS system is the large number of uniform color scales and arrays that are available. What can be done with them in art and design remains largely to be discovered.

The history of almost three decades of difficult work culminating in the production of the OSA-UCS samples has been reported in detail by Dorothy Nickerson, one of the six experts who worked on the project from its beginning [8.31]. At the start, it was recognized that it might be impossible to achieve a perfectly uniform lattice sampling (based on the geometry of inter-locked cubo-octahedra) of color space, for a fixed background. At about the end of the second decade, after confronting various difficult mathematical issues, the Committee was convinced that it was impossible. They decided, however, to continue the program through the production of samples: "... we will produce the best approximation to such a lattice for a neutral 6/ background [Munsell Value=6] that we can design ... ." The difficulty arose from the fact that a circle of six equally different colors (for example: F, M, R, D, A, and G in Fig. 9.1) that are judged to be perceptually equally different from a color at their center O are separated from each other by a difference that actually exceeds their difference from O [Ref. 4.4, p. 166]. The OSA-UCS samples represent a close approximation of the unachievable ideal [Note 8.1]; it is certainly adequate in applications in art and design wherever equal color differences are of interest.

10. Color Names and Notations and Their Levels of Precision

10.1 The ISCC-NBS Color Names of Materials

It is estimated that an experienced person with normal color vision can distinguish about 10 million different surface colors under optimum viewing conditions and that in commerce about half a million different colors are recognized [Ref. 1.18, p. 388]. These facts show the need for a precise numerical system for color specification, such as $CIE(x, y, Y)$. In many instances, however, great precision is not necessary and several hundred color names will suffice. For such purposes, in the United States an official list of 267 standardized color names that apply to the colors of nonluminous materials was presented jointly by the Inter-Society Color Council and the National Bureau of Standards (ISCC-NBS) [7.2].

In 1933, the ISCC undertook to devise a means to designate color for pharmaceutical use that was "sufficiently standardized as to be acceptable and usable by science, sufficiently broad to be appreciated and used by science, art and industry, and sufficiently commonplace to be understood, at least in a general way, by the whole public" [Ref. 7.2, p. 1]. A system of color names was published in 1939; it was revised in 1955. The present list of *ISCC-NBS color names* is given in Table 10.1. The listed names apply to opaque materials, but they can be extended to include light-transmitting materials by substituting "colorless" for "white", "faint pink" for "pinkish white", "faint yellow" for "yellowish white", etc., as shown in Table 10.1. The color names are numbered; these numbers, called *centroid numbers,* may also be used to designate a color.

The ISCC-NBS color names and centroid numbers refer to well-defined color-name blocks or zones that subdivide the region of Munsell color space that includes most commonly experienced surface colors (Fig. 10.1). Each color-name block defines a range of colors in color space that can be represented by one common characteristic color name. The forms of the individual color-name blocks and their locations in color space can be determined from a set of 31 color-name charts published in *Color: Universal Language and Dictionary of Names* by K.L. Kelly and D.B. Judd [7.2].

The 31 color-name charts are intended for the designation of ISCC-NBS color names. To use the charts, a point is located in Munsell color space, which point represents the color for which the ISCC-NBS color name

Table 10.1. ISCC-NBS names and centroid numbers for colors of opaque, translucent, and transparent materials. Centroid numbers 9, 92, 153, 189, 231, and 263 represent different names (in parentheses) for colors of translucent and transparent materials. The tabulation is subdivided into sections by hue names and hue-name abbreviations. This table is based on [Ref. 1.38, p. 448]

Neutral colors: white (Wh) gray (Gy), black (Bk)	37 moderate rO	77 moderate yBr
263 white (colorless)	38 dark rO	78 dark yBr
264 light Gy	39 grayish rO	79 light grayish yBr
265 medium Gy		80 grayish yBr
266 dark Gy	Reddish brown (rBr)	81 dark grayish yBr
267 black	40 strong rBr	
	41 deep rBr	Yellow (Y)
	42 light rBr	82 vivid Y
Pink (Pk)	43 moderate rBr	83 brilliant Y
1 vivid Pk	44 dark rBr	84 strong Y
2 strong Pk	45 light grayish rBr	85 deep Y
3 deep Pk	46 grayish rBr	86 light Y
4 light Pk	47 dark grayish rBr	87 moderate Y
5 moderate Pk		88 dark Y
6 dark Pk		89 pale Y
7 pale Pk	Orange (O)	90 grayish Y
8 grayish Pk	48 vivid O	91 dark grayish Y
9 pinkish Wh, (faint Pk)	49 brilliant O	92 yellowish Wh (faint Y)
10 pinkish Gy	50 strong O	93 yellowish Gy
	51 deep O	
Red (R)	52 light O	Olive brown (OlBr)
11 vivid R	53 moderate O	94 light OlBr
12 strong R	54 brownish O	95 moderate OlBr
13 deep R		96 dark OlBr
14 very deep R		
15 moderate R	Brown (Br)	Greenish yellow (gY)
16 dark R	55 strong Br	97 vivid gY
17 very dark R	56 deep Br	98 brilliant gY
18 light grayish R	57 light Br	99 strong gY
19 grayish R	58 moderate Br	100 deep gY
20 dark grayish R	59 dark Br	101 light gY
21 blackish R	60 light grayish Br	102 moderate gY
22 reddish Gy	61 grayish Br	103 dark gY
23 dark reddish Gy	62 dark gryish Br	104 pale gY
24 reddish Bk	63 light brownish Gy	105 grayish gY
	64 brownish Gy	
	65 brownish Bk	Olive (Ol)
Yellowish pink (yPk)		106 light Ol
25 vivid yPk		107 moderate Ol
26 strong yPk	Orange yellow (OY)	108 dark Ol
27 deep yPk	66 vivid OY	109 light grayish Ol
28 light yPk	67 brilliant OY	110 grayish Ol
29 moderate yPk	68 strong OY	111 dark grayish Ol
30 dark yPk	69 deep OY	112 light Ol Gy
31 pale yPk	70 light OY	113 Ol Gy
32 grayish yPk	71 moderate OY	114 Ol Bk
33 brownish pink	72 dark OY	
	73 pale OY	Yellow green (YG)
Reddish orange (rO)		115 vivid YG
34 vivid rO	Yellowish brown (yBr)	116 brilliant YG
35 strong rO	74 strong yBr	117 strong YG
36 deep rO	75 deep yBr	
	76 light yBr	

118 deep YG
119 light YG
120 moderate YG
121 pale YG
122 grayish YG

Olive green (OlG)
123 strong OlG
124 deep OlG
125 moderate OlG
126 dark OlG
127 grayish OlG
128 dark grayish OlG

Yellowish green (yG)
129 vivid yG
130 brilliant yG
131 strong yG
132 deep yG
133 very deep yG
134 very light yG
135 light yG
136 moderate yG
137 dark yG
138 very dark yG

Green (G)
139 vivid G
140 brilliant G
141 strong G
142 deep G
143 very light G
144 light G
145 moderate G
146 dark G
147 very dark G
148 very pale G
149 pale G
150 grayish G
151 dark grayish G
152 blackish G
153 greenish Wh (faint G)
154 light greenish Gy
155 greenish Gy
156 dark greenish Gy
157 greenish Bk

Bluish green (bG)
158 vivid bG
159 brilliant bG
160 strong bG
161 deep bG
162 very light bG
163 light bG
164 moderate bG
165 dark bG
166 very dark bG

Greenish blue (gB)
167 vivid gB
168 brilliant gB
169 strong gB
170 deep gB
171 very light gB
172 light gB
173 moderate gB
174 dark gB
175 very dark gB

Blue (B)
176 vivid B
177 brilliant B
178 strong B
179 deep B
180 very light B
181 light B
182 moderate B
183 dark B
184 very pale B
185 pale B
186 grayish B
187 dark grayish B
188 blackish B
189 bluish Wh (faint B)
190 light bluish Gy
191 bluish Gy
192 dark bluish Gy
193 bluish Bk

Purplish blue (pB)
194 vivid pB
195 brilliant pB
196 strong pB
197 deep pB
198 very light pB
199 light pB
200 moderate pB
201 dark pB
202 very pale pB
203 pale pB
204 grayish pB

Violet (V)
205 vivid V
206 brilliant V
207 strong V
208 deep V
209 very light V
210 light V
211 moderate V
212 dark V
213 very pale V
214 pale V
215 grayish V

Purple (P)
216 vivid P
217 brilliant P
218 strong P
219 deep P
220 very deep P
221 very light P
222 light P
223 moderate P
224 dark P
225 very dark P
226 very pale P
227 pale P
228 grayish P
229 dark grayish P
230 blackish P
231 purplish Wh (faint P)
232 light purplish Gy
233 purplish Gy
234 dark purplish Gy
235 purplish Bk

Reddish purple (rP)
236 vivid rP
237 strong rP
238 deep rP
239 very deep rP
240 light rP
241 moderate rP
242 dark rP
243 very dark rP
244 pale rP
245 grayish rP

Purplish pink (pPk)
246 brilliant pPk
247 strong pPk
248 deep pPk
249 light pPk
250 moderate pPk
251 dark pPk
252 pale pPk
253 grayish pPk

Purplish red (pR)
254 vivid pR
255 strong pR
256 deep pR
257 very deep pR
258 moderate pR
259 dark pR
260 very dark pR
261 light grayish pR
262 grayish pR

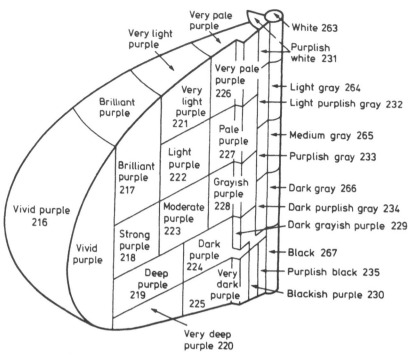

Fig. 10.1. ISCC-NBS color names and centroid numbers. The color-name blocks that partition the purple sector $(3P–9P)$ of commonly experienced colors in Munsell color space. A color name and centroid number has been assigned to each block. (Modification of [Ref. 7.2, Fig. 6])

is desired. The color-name block that contains the point is labeled with the color name sought and the corresponding centroid number.

An alternative procedure is provided by this book. It makes use of Tables 10.1 and 12.2, which present in abbreviated tabular form the information given in graphical form in the 31 color-name charts of [7.2]. The method for using Table 12.2, which leads to the centroid number, is described in Sect. 12.3. The centroid number can then be used to locate the ISCC-NBS color name in Table 10.1.

The ISCC-NBS color names have been adopted by *Webster's Third New International Dictionary* (Unabridged) [1.38]. But a color name can be assigned with certainty only if the Munsell specification is known. If a color specification is given in $\text{CIE}(x, y, Y)$ notation, conversion to the specification in Munsell notation can be made by use of the set of charts and table in Sect. 12.2. In Table 8.1, several color-sample systems are indicated [for example, the Swedish *SIS Colour Atlas NCS*, the *Color Atlas* of Hickethier, and the German standard (DIN-6164)] that include conversions from their notations to the Munsell notation or to $\text{CIE}(x, y, Y)$. For some systems in which such conversions are not included, references cited in Table 8.1 tell where such conversions may be found.

Abbreviations for the ISCC-NBS color names are sometimes used. Each color name consists of a term that designates hue and a hue-modifying term that indicates the magnitude of Munsell Value and Chroma (Fig. 10.2). The abbreviations relate to the two terms. The hue terms with their abbreviations (which are without punctuation) appear as the 26 section titles in Table 10.1, for example, red (R), brown (Br), orange yellow (OY), yellowish brown (yBr), black (Bk). The hue-modifying terms (abbreviations with punctuation) are, for example, very pale (v.p.), light (l.), dark (d.), deep (dp.), very deep (v.dp.), brilliant (brill.), dark grayish (d.gy.), and black-

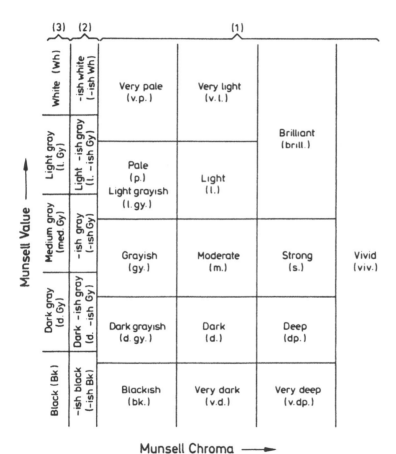

Fig. 10.2. Scheme for generating ISCC-NBS color names (Table 10.1) and their abbreviations [Note 10.1]. (*1*) Hue-modifying terms used in color names; (*2*) the "-ish" grays; (*3*) the neutral colors. In the diagram, the hue-modifying terms are indicated in relation to their Munsell Value and Chroma. Hue-modifying terms [e.g., vivid (viv.)] are combined with hue terms [e.g., green (G)] to form many color names [e.g., vivid green (viv. G)]. The hue terms and their abbreviations are listed as section titles in Table 10.1. The names for low-Chroma "-ish" colors [e.g., pinkish gray (pk. Gy)] are formed from a hue term [e.g., pink (Pk), hence pinkish (pk.)], and a neutral color name [e.g., gray (Gy)]. (Based on [Refs. 1.38, Fig. 1, p. 448; 7.2, Fig. 11])

ish (bk.) [Note 10.1]. Thus some color-name abbreviations, combining the hue and hue-modifying terms, are; dp.R, l.Br, brill.OY, d.gy.yBr. Figure 10.2 shows typical color-name blocks drawn on a graph of Munsell Chroma versus Munsell Value. For the off-neutral gray series (from black Bk to white Wh), with "-ish" modifying terms, typical abbreviations are: br.Gy for brownish gray (64), pk.Wh for pinkish white (9), and p.Bk for purplish black (235).

10.2 The ISCC-NBS Centroid Color Charts

A set of 267 glossy painted samples (2.5 × 2.5 cm) whose specifications correspond to the central points of the color-name blocks in Munsell color space has been produced by the National Bureau of Standards. This sample set is provided on charts, the *ISCC-NBS Centroid Color Charts* [10.1]. The background of the samples is neutral gray, but the lightness of the gray varies across the chart such that "each color is seen on a background of approximately its own lightness." The centroid number and the abbreviation of the ISCC-NBS color name identify each sample. A separate table lists the Munsell notation for each sample. The samples were produced to stringent tolerances and constitute "an extremely useful and inexpensive set of special color standards" [Ref. 7.2, p. A-10].

10.3 The Dictionary of Color Names (ISCC-NBS)

In some color-sample systems, color names such as "bambino", "Bavarian blue" and "Hooker's green" are given to specific color samples. If a color-sample system is one of the 14 listed in [Ref. 7.2, p. 36], the ISCC-NBS color name and centroid number that correspond to the system's color name for a sample (e.g., "Bavarian blue") can be found by consulting the *Dictionary of Color Names* [Ref. 7.2, pp. 85–158] where 7500 color terms are listed. Also indicated, in each case, is the name of the color-sample system that employs the name. Thus, for "Bavarian blue", the corresponding ISCC-NBS color name is "dark blue", the centroid number is 183 and the source is the Plochere Color System (interior decoration) (Table 8.1). The *Dictionary* indicates that "Hooker's green" is included in the color system of Maerz and Paul [*Dictionary of Color* (Table 8.1)] and that two color samples are assigned that name: "Hooker's green No. 1" for which the ISCC-NBS color name and centroid number are "strong yellow green" (131) and "Hooker's green No. 2" for which they are "moderate green" (145).

A tabulation of synonymous names is also provided in [Ref. 7.2, pp. 37–82]. Under each centroid number are listed the various names assigned in the 14 systems to the same, or approximately the same, color. Listed under centroid number 183 ("dark blue") there are 91 different names. If we are interested in the names given to this color in the textile industry, we should consider the six names listed in one of the 14 color-sample systems – namely, that of Textile Color Card Association [now the Color Association of the United States (Table 8.1)]. Listed under centroid number 145 ("moderate green") there are 25 names that appear in the Maerz and Paul system. These names include, in addition to "Hooker's green No. 2", the following that are familiar to artists: "Egyptian green" and "transparent chromium oxide" for two different color samples; "chrome green", "emeraude", "(French) Veronese green", and "viridian" for one color sample.

10.4 Color Designation: The Levels of Precision

After considering the discussions of color specification in this book, we cannot escape realizing that certain systems are more suitable than others in a given situation. Perhaps the most important aspect that must be recognized in selecting a system is the level of precision that is needed. Kelly has described the following six levels of precision of color designation [Ref. 7.2, p. A-10]:

First Level. The first level of precision pertains to the most general color designations, such as the color yellow when one casually indicates the color of a car. At this level, 10 generic hue names suffice: pink, red, orange, brown, yellow, olive, yellow green, green, blue, and purple. In addition, there are the neutrals: white, gray, and black. At this level, Munsell color space is divided into 13 large blocks.

Second Level. At the second level, the list of names is increased by adding 16 intermediate hue names: yellowish pink, reddish orange, reddish brown, orange yellow, yellowish brown, olive brown, greenish yellow, olive green, yellowish green, bluish green, greenish blue, purplish blue, violet, reddish purple, purplish pink, and purplish red. Twenty-six of the colors (omitting white, gray, and black) are presented as the section titles in Table 10.1. At this level of precision, the color of the car might be said to be "greenish yellow". At this level, Munsell color space is sectioned into 29 color-name blocks.

The color names for lights proposed by Kelly (Sect. 7.1) are at this level. The list of 26 color names in Table 10.1 and the list for lights in Table 7.1 are similar. Among those for lights we find, however, orange pink, blue green, and red purple; necessarily absent are brown, olive, gray, and black. The color names for lights have not been revised since their introduction (1939); the ISCC-NBS color names for materials were revised in 1955.

Third Level. At the third level, the aforementioned color categories are subdivided (Table 10.1), yielding the 267 ISCC-NBS color names. At this level, Munsell color space is divided into 267 color-name blocks.

The color names are generated from those at the second level by the addition of hue-modifying terms that designate the degree of Munsell Value and Chroma (Fig. 10.2). These terms include vivid, brilliant, strong, deep, very deep, very light, light, moderate, dark, very dark, very pale, pale, light grayish, grayish, dark grayish, and blackish. At this level of precision, the color of the car might be found to be "vivid greenish yellow" (centroid number 97). The samples of the *Standard Color Reference of America* and of the *Horticultural Colour Chart* (Table 8.1) belong at this level.

Fourth Level. At the fourth level, color space is further subdivided into from 1000 to 10000 colors. The color-sample systems found at this level of precision include the *Munsell Book of Color*, the *SIS Colour Atlas NCS*, the *Colour Harmony Manual*, *Chroma Cosmos 5000*, the *Plochere Color System*, the *DIN-Farbenkarte*, and *The Dictionary of Color* by Maerz and Paul (Table 8.1). It is possible, but rather cumbersome, to employ color names at this level. Numerical or letter codes are preferable, such as the Munsell notation. With the aid of the *Munsell Book of Color*, glossy edition, the color of the car's paint might be found to be in Munsell notation, 7.5Y 8/12.

Fifth Level. The fifth level is demonstrated by visually interpolating between Munsell chips. By this means, the number of specified colors can be increased to about 100000. This is possible, because, with great care, Value can be estimated to 1/10th of a Munsell Value unit, Chroma to 1/4th of a Chroma unit, and Hue to 1 Hue unit at Chroma 2 or to 1/4 Hue unit at Chroma 10 and higher (usually 1/2 Hue unit is used). Thus, the color of the paint on the car might be found to be $8\frac{1}{2}$Y 8.3/$12\frac{1}{2}$.

Sixth Level. Finally, at the sixth level, optical instrumentation is required to provide a measurement. Here, the number of color divisions is of the order of 5000000. The color of the car paint, now specified by a laboratory, might be CIE 1931 (0.291, 0.433, 0.468), CIE ILL C. These specifications can then be converted to Munsell notation by the use of charts (Sect. 12.2) to comparable accuracy [8.20]: 8.6Y 8.3$_5$/12.6.

11. Conditions of Viewing and the Colors Seen

11.1 Psychological Aspects and Color Systems

Color is a major subjective topic in the domain of psychology. In color-vision research, systematic studies are commonly made on human visual response. In such studies, sensory scales are devised to assess hue, saturation, brightness, brilliance, color difference, etc. Objective physical measures cannot be used; they are not applicable.

Important developments in color science stem from the linkage of subjective color response to objective physical stimuli. Colorimetry, in particular, has become a highly precise quantitative science. However, in order to provide a rational basis to which objective measures could be applied, it was decided to adopt a psychophysical definition of light and color (Sect. 6.3).

In the research leading to the $CIE(x, y, Y)$ system of color designation, basic data were obtained by physically measuring the amount (or reception rate) of light required to produce matches of visual color response while viewing, in dark surrounds, color patches produced by light beams. The data obtained were transformed to produce the tristimulus (color-matching) curves for the CIE imaginary primaries (Fig. 6.8). Although the system is a psychophysical tool of colorimetry, we should not overlook the fact that the tristimulus curves of the CIE 1931 standard observer and the characteristics of the light that reaches "his" eyes determine "his" color response under particular, yet realistic, conditions. This response is described by the roughly approximate, but very useful correlates to perceived hue, saturation, and brightness (or lightness) – namely, dominant (or complementary) wavelength, purity, and luminance (or luminance factor), respectively.

We have seen that the Munsell system offers samples of colors whose Hue, Chroma, and Value vary approximately in accordance with perceptually uniform scales [Note 8.1]. In the Swedish Natural Colour System (NCS), the hues are differentiated perceptually with respect to their binary composition of unitary hues; NCS chromaticness, NCS whiteness, and NCS blackness are established visually. The German DIN system incorporates the parameter DIN-Sättigung, scaled on the basis of perceived saturation comparisons. Specifications for the OSA uniform color scales were devised with the use of relations established through experimental color-difference measurements [Ref. 4.4, p. 165].

Often these systems are used in applications for which they were not intended, with satisfactory results. Sometimes, however, more attention should have been paid to the standard conditions that are carefully specified for their use. In the case of the Munsell color chips, for example, valid comparison with a test sample can be made only when both the Munsell chip and the test sample rest on a neutral background (middle gray to white) and daylight illumination (or its equivalent) is provided. The viewer should view the surface at 90° with the illumination striking the surface at 45° [8.19].

The colors of objects surrounding us are often perceived under conditions that differ from those prescribed in the above systems. A dab of colored paint in a painting is not usually found surrounded by dabs of neutral gray paint. It is not surprising that the perceived color of the paint might differ importantly from its color perceived on a neutral gray background. And its perceived color might continue to vary in unexpected ways, if the eye, roving from one dab to another, is influenced by afterimages or if the painting is viewed in different illuminations. Such variations in color response are normal in our everyday experience with color vision. As we shall see, the origin of such variations may be physiological as well as psychological. In this chapter, several aspects of color are discussed that relate to topics introduced earlier in this book.

11.2 Hue Responses

The subject of hue responses is an interesting and helpful introduction to the relevant topics of chromatic adaptation, afterimages, and simultaneous contrast. Jameson and Hurvich have introduced this approach in their research on color vision within the framework of a quantitative opponent-colors model that is based on Ewald Hering's original theory of color vision (1878), in which the psychological primaries play a central role (Sects. 3.2, 8.7). Some of their earlier experiments made use primarily of colors produced by spectral lights (lights of single wavelength) [Refs. 2.3, p. 55; 3.6].

Jameson and Hurvich measured redness, yellowness, greenness, blueness, and whiteness in *binary hues* produced over the full wavelength range, 400–700 nm, under conditions of neutral adaptation (that is, for the situation where the eyes are adapted to daylight). For example, the perceived yellowness of 510-nm light (yellowish green) was determined by mixing a given amount of the light with increasing amounts of 475-nm light (unitary blue) until the yellow was "canceled" chromatically, leaving unitary green. The relative amount of unitary blue required was taken as a measure of the amount of yellowness produced by 510-nm light. The experiment was repeated for 520-nm light, 530-nm light, and so on, in 10-nm intervals, to 700-nm light – that is, the wavelength range of spectral light that produces yellow-containing colors. Greenness, blueness, and redness were measured in essentially the same manner.

The results presented as curves called chromatic response functions varied somewhat between the two observers. Those for one observer, however, correspond closely with those predicted by calculation for the CIE 1931 standard observer (Plate XIV). One chromatic response function applies both to yellowness (arbitrarily plotted above the horizontal axis) and to blueness (below the horizontal axis); the other, both to redness (above the horizontal axis) and to greenness (below the horizontal axis). The third line (black) applies to whiteness; it is taken to be equivalent to that of relative luminosity (Sect. 4.7, and curve II in Fig. 6.8) [Ref. 2.3, p. 62]. A blackness response function is not shown; there is no way of directly varying a stimulus to produce blackness.

Yellow and blue are *opponent hues;* in mixture, yellow and blue lights "cancel" chromatically producing a hueless response (white) or a less-saturated yellow or blue response (no bluish yellows or yellowish blues are possible). This opponency is indicated by arbitrarily assigning positive color response values to yellowness and negative values to blueness. Because red and green are opponent hues, too, a similar assignment of positive values to redness and negative values to greenness is also made.

Given the chromatic response functions, we can now read from them the hue compositions (in relative amounts of response) of any spectral light. Thus, at 460 nm (Plate XIV), we find, by reading the scale on the vertical axis, the relative amounts (neglecting the minus sign): red (0.23), blue (0.64), and white (0.06). Yellow and green are not present, and they cannot be if their opponents red and blue are present. At 477 nm (unitary blue, for the 1931 standard observer), the composition is given by the following relative amounts: blue (0.34) and white (0.13). Yellow, being the opponent of blue, is not present, nor is the opposing pair red and green.

The chromatic response functions shown in Plate XIV apply at moderate level of spectral light intensity and to neutral adaptation. If the intensity of the spectral light is changed sufficiently, the perceived hue changes (except in the case of the four unitary hues, which do not vary with light intensity). This phenomenon is called the *Bezold-Brücke effect* [Refs. 1.20, p. 54; 2.3, p. 72]. Specifically, colors of spectral lights that appear red yellow (orange) or green yellow at moderate intensities appear yellower at higher intensities, and wavelengths that appear red blue (purple) or green blue at moderate intensities appear bluer when the luminance level is increased.

According to [Ref. 2.3, p. 80], the unitary hues can be produced by single-wavelength light at 578, 499, and 477 nm, respectively, for the CIE 1931 standard observer (Fig. 11.2). Among people with normal color vision, the specific wavelengths have been found to vary within the following ranges: 568 to 583 nm (unitary yellow), 495 to 535 nm (unitary green), and 467 to 485 nm (unitary blue) [Ref. 1.20, p. 57]. For the average person, unitary red can be said to be produced by a mixture of single-wavelength light of 400 and 700 nm, for which the complementary wavelength (relative to CIE ILL C) is 495 nm. (Kelly has proposed a complementary wavelength of 493 nm [4.1].)

The Hue circle of the Munsell color system was designed so that the five equally spaced major Hues (5R, 5Y, 5G, 5B, and 5P) would be close to unitary red, yellow, green, and blue and a purple [11.1]. This is achieved within one or two Hue units for NCS unitary red, yellow, and green, and within five Hue units for NCS unitary blue. Table 11.1 shows approximate Munsell Hue number designations (at Value $V = 5$) for the unitary hues and for six intermediate hues [Note 11.1].

Table 11.1. Approximate Munsell Hue equivalents (at $V = 5$) of the NCS unitary hues and their intermediate hues (Fig. 8.23)

NCS unitary hues and intermediate hues	Munsell Hue number	NCS unitary hues and intermediate hues	Munsell Hue number
Unitary red	4	B50G	57
Y50R	15	Unitary blue	70
Unitary yellow	26	R75B	77
G50Y	35	R50B	87
Unitary green	46	R25B	97

11.3 A Psychological Color Specification System

Unlike other color systems considered earlier in this book, the Hue, Brightness, Saturation (HBS) system devised by Hurvich and Jameson specifies quantitatively the perceived colors of lights and of objects by three psychological parameters that are correlates of three basic perception attributes [Refs. 2.3, Chap. 7; 3.6; 11.2]. The three parameters are *hue coefficient, saturation coefficient,* and *metric lightness.* Much like the CIE (x, y, Y) system, it can serve in color specification without the aid of material color standards.

The hue coefficient is taken to represent the relative amount of one hue contained in a binary hue divided by the sum of the relative amounts of both hues. The relative amounts of the hues can be read from the chromatic response functions plotted in Plate XIV. For the 460-nm spectral light, the hue coefficient 0.74 (or 74 %) for blue is obtained by dividing 0.64 (blue) by the sum of 0.64 (blue) and 0.23 (red). For 477-nm light (unitary blue, only) the hue coefficient is 1.00 (or 100 %).

The saturation coefficient for spectral lights is calculated by dividing the sum of the relative amounts of both hues in a binary hue by the sum of the relative amounts of both hues and the white component. Thus, for

460-nm spectral light, the saturation coefficient of 0.94 (or 94%) was obtained by dividing the sum of the relative amounts of the hues (0.87) by the sum of the relative amounts of the hues and of the white component (0.87 + 0.06). The saturation coefficient 0.72 (72%) for the blue (477 nm) is obtained by dividing the relative amount of blue (0.34) by the sum of the relative amounts of blue and white (0.34 + 0.13). Hurvich and Jameson claim that "any real stimulus that elicits a chromatic response always elicits a simultaneous response in the achromatic visual process" [11.2]. This means that neither the saturation coefficient nor perceived saturation can be 100% (Sect. 4.3).

Determination of the hue and saturation coefficients for the colors of illuminated objects involves a procedure much like the one employed in determining the CIE tristimulus values. In the HBS case, the chromatic response functions are used (Plate XIV) [Ref. 2.3, Chap. 7] [Note 11.2]; whereas the color-matching functions, or tables based on them, are used in the CIE case (Fig. 6.8). After citing supporting data obtained by other investigators, Hurvich and Jameson stated that they "consider the theoretically derived hue and saturation expressions sufficiently valid to serve as a basis for a system of color specification in terms of the psychological attributes of the color sensations" [11.2].

The third parameter, metric lightness L^* [Note 7.3], is currently used for the colors of nonluminous and luminous [7.35] objects. This parameter may have to be modified when means are developed to accommodate color-contrast effects in the HBS system [11.3].

In HBS color space, brightness (for lights) or lightness (for objects) is measured in equal perceptual steps along the vertical axis. The horizontal plane displays a circular psychological color diagram or *HBS color diagram*. The equally spaced concentric circles represent lines of constant saturation coefficient S (Fig. 11.1). Radial lines from the center are lines of constant hue coefficient. The hue circle is divided into four quadrants, with unitary red, yellow, green, and blue (indicated on the outer circumference by 1.00R, 1.00Y, 1.00G, and 1.00B) at the 12, 3, 6, and 9 o'clock positions, respectively. Each quadrant is subdivided into equal steps of hue coefficient. Equal angles between radii represent equal hue differences and equal radial distances from the center represent equal saturations [11.2]. The hue circle is equivalent to the Hering hue circle shown in Plate X. All of the binary red-yellow hues are contained in quadrant I, between 1.00R and 1.00Y; all of the red-blue hues are in quadrant II; all of the blue-green hues are in quadrant III; all of the green-yellow hues are in quadrant IV.

Points that represent spectral lights at various wavelength intervals from 400 nm to 650 nm are shown in Fig. 11.2. The curve drawn through the points is the spectrum locus in the HBS color diagram. (The HBS color diagram is not a chromaticity diagram.) The variations of hue and saturation differences between colors that differ equally in wavelength reflect variations

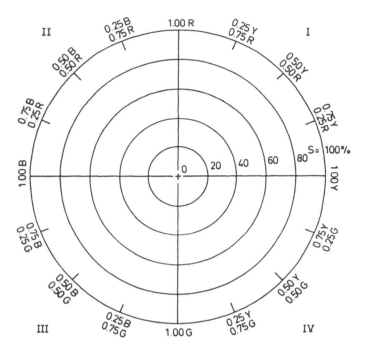

Fig. 11.1. HBS color diagram Concentric circles are lines of constant saturation coefficient S ($S = 0, 20, 40, \ldots, 100\%$). Any radial line is a line of constant hue coefficient. Four radial lines that correspond to the unitary hues (1.00R, 1.00Y, 1.00G, 1.00B) divide the diagram into four quadrants (I, II, III, IV). Intermediate hues are identified by their binary compositions

of the sensitivity of the visual system to wavelength changes and, hence, the ability of the eye to discriminate wavelengths in different ranges [11.2].

Figure 11.2 contains four points (within circles) that represent the hues and saturations of colors produced by four objects that have equal lightness (CIE ILL C): object 1 (magenta with a slight amount of blue), object 2 (cyan), object 3 (green yellow), object 4 (reddish yellow) [11.2]. CIE ILL C is represented by the central point.

The HBS color diagram can be used to represent hue and saturation under various states of chromatic adaptation (Sect. 11.4). However, a set of chromatic response functions must be provided for each state of adaptation before the necessary calculations can be performed.

The HBS hue circle is structurally the same as the Swedish NCS hue circle. The HBS hue coefficient and the NCS hue designation are similar; both report binary hues by their relative hue compositions. In the NCS, the compositions are judged visually; in the HBS scheme they are determined from chromatic response functions. In the NCS, equal steps or intervals along the hue circle do not designate perceptually equal hue differences. (According to G. Tonnquist, it has been shown that "although one quadrant (e.g., the

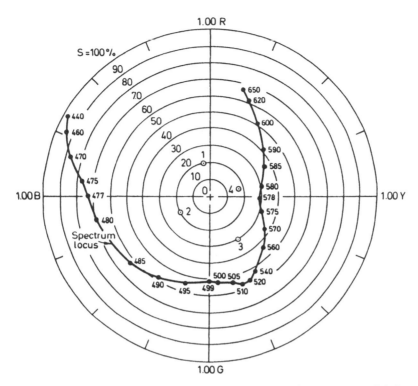

Fig. 11.2. HBS color diagram showing the spectrum locus and points (•) that represent the HBS hue and saturation coefficients of spectral lights (wavelength [nm]). (Based on data from [Ref. 2.3, p. 80].) o: Four object colors of constant lightness (Sect. 11.3). Neutral adaptation, CIE ILL C

blue-red) contains more distinguishable hue steps than another (e.g., the yellow-green) we can obtain no more color combinations of fundamentally different character with the hues from any particular quadrant than with the hues from any other quadrant" [8.26, see also 8.49].) On the other hand, as stated above, in the HBS hue circle, equal angles between radii represent equal differences of perceived hue [11.2]. Further experimental study may be required to resolve this apparent disagreement. Although hue designation is similar in the two systems, they differ in the two other perceptual variables that they employ.

The HBS system has not been adopted generally for the specification of color. The CIE (x, y, Y) system has served very well in this function in widespread industrial applications. But with increasing interest turning to the diverse psychological effects commonly experienced in color vision and to the broad subject of color appearance, the HBS or a similar system may prove to be useful.

11.4 Adaptation to Color; Color Constancy

A visual experience that we frequently have, sometimes rather dramatically, is *dark and light adaptation*. When we step from bright sunlight out of doors into a dimly lit room, we suddenly feel immersed in blackness. Twenty or thirty minutes, or perhaps an hour, may pass before our visual system becomes fully adapted to the low level of illumination. When we return to the out-of-doors, our eyes are instantly dazzled by the bright sunlight, but we adapt relatively quickly to it (in about one minute) [Ref. 2.5, p. 19]. The change of sensitivity of the retina in adapting to the change of level of illumination is of the order of 1000 to 1 [11.4]. The illumination levels for very bright sunlight and for a dimly lit room are about 100 000 and 10 illuminance units (lux), respectively [Ref. 7.23, p. 587]. The illumination level may be as high as 200 in a well-lit living room. On a very dull day, the illuminance from the sky may be as low as 100 lux and moonlight illumination may be as high as 0.1 lux. The lower limit of photopic vision occurs at illumination of about 1 lux (Sect. 2.2) [2.7, Fig. 2].

We also experience adaptation to color (*chromatic adaptation*), but in a more subtle, yet significant, way. The changes of retinal sensitivity are the response to changes of the chromatic qualities of the light rather than to changes of the level of illumination [Ref. 2.3, p. 195]. Consider what we see when we pass from a room bathed in daylight into a room well illuminated by an incandescent-tungsten-filament light bulb. On entering the second room, we see the colors affected by the faint yellowish orange light in comparison with their appearance in daylight. After a few minutes when the eyes have completely adjusted in sensitivity to the light provided by the light bulb, the colors of the objects appear more like, but usually not exactly like, their colors perceived in daylight [Ref. 1.18, p. 354].

A theoretical study has been made of the shift of perceived colors of objects when the illumination is changed from daylight (CIE ILL C) to that provided by an incandescent lamp (CIE ILL A). For example, a bright magenta (with a small blue component) object color was considered in daylight (CIE ILL C) with the CIE 1931 standard observer's eye adapted to daylight [11.2]. The illumination was assumed to change suddenly to incandescent lamplight (CIE ILL A), and the color "perceived" was found to be yellow red before chromatic adaptation had a chance to occur. After adaptation to the lamplight, the color "perceived" was a pure (almost unitary) red. These predicted changes of color are given in Table 11.2 along with those for three other brightly colored objects. (HBS designations for the colors are given in [11.2].)

The initially perceived colors of the four objects are represented (open circles) in Fig. 11.2. They are shown again at a larger scale in Fig. 11.3 and with subsequent changes indicated following a change of illumination. The results for unchanged adaptation to CIE ILL C are represented by triangles;

Table 11.2. Changes of object color with a change of illumination

Time	Illuminant	Adaptation state of the eye	Perceived colors			
			Object 1	Object 2	Object 3	Object 4
1) 0	CIE ILL C	CIE ILL C	Magenta	Green blue	Yellow green	Reddish yellow
2) 1 sec. later	CIE ILL A	CIE ILL C	Yellow red	Yellow green	Orange yellow	More reddish
3) 5 min. later	CIE ILL A	CIE ILL A	Pure red (increased saturation)	Blue green (slightly increased saturation)	Yellow green (decreased saturation)	Reddish yellow (slightly increased saturation)

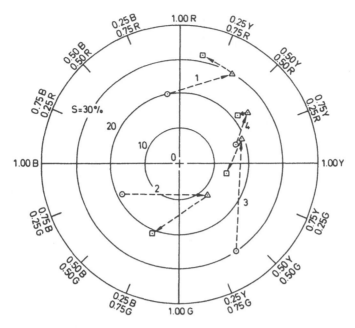

Fig. 11.3. HBS color diagram. Illustration of chromatic adaptation. The illumination of four objects is changed from CIE ILL C to CIE ILL A (Table 11.2). ○: Initial state, illumination and adaptation CIE ILL C; △: intermediate state, illumination CIE ILL A and adaptation CIE ILL C; □: final state, illumination and adaptation CIE ILL A. CIE 1931 standard observer. (Based on [Ref. 11.2, Fig. 6])

the results for adaptation to CIE ILL A are shown by squares. The first leg of the bent path (from circle to triangle) shown for object 1 represents the initial change of color perceived (from magenta to yellow red) when the illumination is suddenly changed from CIE ILL C to CIE ILL A and chromatic adaptation to CIE ILL A has not had sufficient time to occur. But, after the eye adapts to CIE ILL A, the perceived hue returns at least part of the way back to the original color, as shown by the second leg of the bent path (from triangle to square).

The bent path illustrates the tendency of the visual system to minimize differences in perceived color when the illumination is changed. This phenomenon is called *color constancy*. Some people prefer to call it approximate color constancy, because there is almost always some remaining difference in perceived chromatic colors when the illumination is changed. Objects 1 and 2 clearly display a greater degree of color constancy than object 3, and object 4 displays the greatest degree.

White and neutral gray surfaces tend to show almost complete color constancy in the two illuminations (CIE ILL C and CIE ILL A). It is interesting to note a marked difference between color vision and color photography. Color photographic film, unlike the eye, is unable to change its sensitivity

190

[Refs. 4.4, p. 106; 7.44, p. 214]. A photograph of a sheet of white paper taken in daylight with outdoor color film is white. But a photograph of the same sheet of paper obtained with the same film, but taken indoors in illumination provided by an incandescent-tungsten-filament lamp, has a yellowish orange tint. The correct color, white, can be obtained with film that has a color sensitivity appropriate to the illumination of the paper.

Two sets of chromatic response functions for adaptation to CIE ILL C and to CIE ILL A are shown in Plate XV. The set for adaptation to CIE ILL C was used for determination of the points in Fig. 11.3 shown as circles and triangles; the set for adaptation to CIE ILL A was used to determine those shown as squares. Not shown are the whiteness functions, which also vary with adaptation. Incandescent-lamp light has a yellowish orange tinge. If adaptation to such light were 100 % complete, then color constancy would be observed for all color stimuli. That this does not occur can be seen in the two sets of chromatic response functions in Plate XV. For adaptation to CIE ILL C, we find here that a beam of spectral light of about 575 nm produces unitary yellow, showing a response of 0.47 units of yellow and no red, green, or blue (the white content is not available in the plot). The same light (575 nm) produces a greenish yellow response when the visual system is adapted to CIE ILL A because the response shows about 0.20 units of yellow and 0.20 units of green. For a spectral light of 450 nm, however, the change of chromatic response is hardly noticeable. Adaptation to reddish yellow light seems to reflect a change of balance in the color vision system [3.6]. The depression in the yellow-blue response function in the yellow region is particularly marked, and the extension of the blue branch to higher wavelengths is significant. Similarly, the green branch of the red-green response function is larger.

The changes reported above for adaptation to CIE ILL A are represented under more exaggerated conditions in Plate XVI, for which adaptation to spectral light at 580 nm (taken to be unitary yellow) was assumed. The yellow branch of the yellow-blue response function is almost completely depressed. This means that a beam of light of 580 nm appears nearly white; but total adaptation cannot be assumed [11.5]. The whiteness response function (not shown) possesses a depression in the medium- to long-wave spectral region (as compared with the whiteness response function shown in Plate XIV).

Chromatic response functions for other exaggerated conditions of adaptation (unitary blue, green, and red) are also shown in Plate XVI. In each case, adaptation to an illumination that produced a unitary hue resulted in a depression of the response to that hue and an increase of response to the opponent hue. In the companion chromatic response function, the balance between opponent hues is relatively unchanged.

The HBS color diagram in Fig. 11.4 shows the predicted perceived color change for the spectral light (580 nm) from white adaptation (circle) to yel-

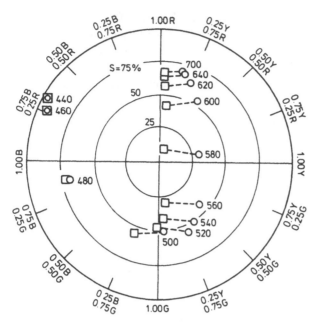

Fig. 11.4. HBS color diagram with points that represent the HBS hue and saturation coefficients of spectral lights (wavelength [nm]) seen under conditions of neutral (white) adaptation (o) and adaptation to spectral light (yellow, 580 nm) (□). CIE 1931 standard observer. (Based on [Ref. 11.2, Fig. 3])

low (580 nm) adaptation (square). The location of the square shows the perceived color to be nearly white (the saturation coefficient is approximately 10 %). Also shown is the general shift away from yellow towards blue of the chromatic response to other spectral lights. But even under these exaggerated conditions of adaptation change, the perceived changes for stimuli at 480, 460, and 440 nm are minor. Data points on HBS color diagrams of the shifts in response occurring for stimuli when adaptation is shifted from neutral to unitary blue, green, and red adaptation would show trends similar to those above – that is, a general shift away from blue towards yellow after adaptation to blue; away from green towards red after adaptation to green; and away from red towards green after adaptation to red [11.2].

If the eye turns from neutral illumination to one that produces a binary hue (instead of a unitary hue), then, on adaptation, there will be changes of the balance between opponent hues in both chromatic response functions [11.5]. The case of CIE ILL A discussed above is an illustration of this (Plate XV).

We may ask, "What is the change of perceived color of a patch of paint when it is viewed in two different illuminations, say CIE ILL C and CIE ILL A;" On a HBS color diagram, such as Fig. 11.3, the changes of hue and saturation coefficients can be seen from the positions of the circles and squares, and the change of metric lightness can be calculated. The basis for these calculations has been discussed in Sect. 11.3. In present-day commerce, industry, and science, the CIELAB and CIELUV formulas, which are based on two of the best available approximations to uniform color space (Sect. 8.2), are used for color-difference determinations [Note 7.4]. The use of these formulas for establishing color differences *under daylight illumination* was discussed briefly in Sect. 7.10. These formulas yield a total color difference. In addition, hue, saturation, and lightness component differences (CIELAB, CIELUV) may be determined: metric hue angle, metric chroma, and metric lightness (Sect. 8.2).

Color-difference data for surfaces illuminated by two sources (usually CIE ILL C and CIE ILL A) are often required in commerce. Given the spectral reflectance curve for a colored surface, one can calculate the values for CIE (x, y, Y) for the surface in one illumination and then in the other (both for adaptation to CIE ILL C). The path in CIE color space (Sect. 8.1) from the first point (x_1, y_1, Y_1) to the second point (x_2, y_2, Y_2) represents the *colorimetric color shift* [Ref. 7.44, p. 212]; it corresponds to the first leg of a bent path projected on a horizontal plane (Fig. 11.3) of HSB color space. A separate calculation is required to take adaptation into account – that is, to establish the *adaptive color shift* (the second leg of the bent path) by determining a third point (x_3, y_3, Y_3), the chromaticity of the *corresponding color* (Sect. 7.12) [Note 7.6]. Then a color-difference formula (CIELUV or CIELAB) is used to calculate the perceived color difference resulting from two states of chromatic adaptation – that is, between the colors represented by (x_1, y_1, Y_1) and (x_3, y_3, Y_3). The use of these formulas, intended for neutral adaptation, is admissible because the difference between the adaptation states is small.

In Sect. 7.12, the color changes of two samples (metameric with respect to CIE ILL C) were discussed when their illumination was changed to CIE ILL A. The changes of colors reported were determined by the above method. If color constancy were complete (100 %), the points in color space (x_1, y_1, Y_1) and (x_3, y_3, Y_3) would be identical, the color difference would be zero, the occurrence of metamerism would not be indicated because there would be no mismatch with the use of the second illuminant, and the color-rendering index of the second lamp would be 100 on the basis of the matching samples [11.6]. As Wright has pointed out, color constancy, metamerism, and color rendering are interrelated problems [Ref. 6.2a, p. 152].

Other explanations for color constancy have been offered. For example, learning that a lemon is yellow may help us to see it with unchanged color when its illumination is changed (Sect. 11.6). Learning may be a contributing factor. But chromatic adaptation, of which color constancy is a result, applies generally, including to unfamiliar objects. In fact, it has been shown that some animals (monkeys, hens) exhibit the effects of chromatic adaptation to a light source [Ref. 7.44, p. 212].

11.5 Adaptation Level

In Sect. 11.4, it is stated that white and neutral gray surfaces tend to display almost complete color constancy in daylight and incandescent-lamp light. However, a white surface can take on a "cold" bluish cast when it is illuminated with north-sky light or a "warm" yellowish cast when it is illuminated by an incandescent-tungsten-filament lamp. Furthermore, perhaps unexpectedly, a neutral dark gray surface may exhibit a yellowish cast in north-sky light and a bluish cast in lamplight [Ref. 2.3, p. 199].

H. Helson reported three experiments that illustrate in a general way the above phenomenon. A number of observers were asked to describe what they saw when they viewed a series of graded neutral gray chips on white, gray, and black backgrounds in strongly chromatic light (a monochromatic red) [Refs. 1.20, p. 89; 11.7, p. 576; 11.8]. On the white background, the gray chip of luminance factor 0.27 appeared achromatic. But lighter gray chips had a red cast, and darker gray chips, a blue green cast. On the gray background, the gray chip that had an achromatic appearance was somewhat darker (luminance factor 0.10). On the black background, the luminance factor of the achromatic-appearing chip was 0.03. In the second and third experiments, as in the first, a red cast was noted on chips of luminance factor higher than 0.10 and 0.03, respectively; and, in the case of the gray background, a blue green cast was observed when the luminance factor was lower.

Generally speaking, the average luminance-factor level of a scene affects the appearance of an object in the same way. This has been termed the *adaptation level* by Helson. Thus, a surface lighter than the surroundings takes on the cast of the hue of the illuminant; a surface darker than the surroundings takes on the cast of the hue of the afterimage complementary (Sect. 11.7) of the illuminant's hue [Refs. 11.8,9; 11.10, Fig. 16]. This phenomenon is known as the *Helson-Judd effect* [Ref. 1.20, p. 88].

11.6 Memory Color

Whenever we associate the color yellow with bananas or lemons and red with ripe tomatoes we are employing memory colors. They are colors associated with familiar objects in daylight, and, as such, they constitute an individual's standard, relatively unchanging references [Refs. 1.20, p. 87; 7.44, p. 215; 11.8,11]. (A color remembered, such as that of a passing car, is, however, not a memory color. Color memory is another topic; it is not considered here.)

C.J. Bartleson has made an experimental study of memory colors associated with a number of familiar objects [11.11]. Fifty observers were asked to select Munsell color samples corresponding to their memory colors. He found, for example, that the memory color of red bricks has more purity than their measured color. The memory colors of sand and soil are more yellow and also of greater purity. In the case of green grass and deciduous foliage, the memory color is more blue green in hue. The memory color of blue sky is more cyan and of higher purity than the actual sky. The memory color for flesh (Caucasian) tends to be more yellow and of higher purity than natural flesh color. Generally, a memory color tends to be of enhanced purity and of a hue shifted "in the direction of what is probably the most chromatic attribute of the object in question."

Memory colors are generally more pleasing to individuals. They are desired in photographs and painted portraits, frequently in preference to faithfully reproduced colors. Memory colors are also said to be capable of influencing color perception [Refs. 1.20, p. 87; 5.18, p. 110]. For example, a banana cut from a piece of yellow paper might be found not to match a noncontiguous square cut from the same paper. In this test, the memory color has influenced the perception of the color of the paper cut in the shape of a banana but not the perception of the color of the paper cut in a neutral shape.

11.7 Afterimage Complementary Color Pairs

Afterimages are frequently occurring visual experiences. When, in a darkened room, we gaze at a lighted light bulb or through a window at the brightly sunlit out-of-doors and then turn our gaze to the dim interior of the room, we continue to see images of the light bulb or of the bright surroundings. Sometimes, for as long as a few minutes, they remain as ghosts drifting before us with the movement of our eyes, often influencing the appearance of objects on which we fix our eyes. Nevertheless, we learn to ignore them, and most of the time we are hardly aware of their presence.

Many types of afterimages are possible [Refs. 1.20, p. 68; 2.3, pp. 183, 186; 11.12]. We shall be concerned, however, only with the common *negative afterimage* of nonluminous objects in which the hues experienced are complementary to those perceived in the original object. Plates XVII–XIX are presented to demonstrate common negative afterimages. The green square produces an afterimage of the complementary hue (magenta) when viewed in daylight with the neutral gray area (below) serving as background. The afterimage viewed against a yellow background has an orange hue, suggesting (simplistically) that a kind of additive mixture is occurring. A similar situation is presented when the afterimage is viewed against a red background, which produces a red square that is significantly more vivid than its surrounds.

Plate XX shows a gray square with green surrounds (the reverse of the situation in Plate XVII). At first we might suppose that no hue would be produced in the afterimage of the square viewed against the gray background. But, surprisingly, a green square afterimage is produced. In a study of this phenomenon, S. Anstis, B. Rogers, and J. Henry have established that two processes contribute to the formation of the green square [11.13]. First of all, while we stare at the gray square, it has a pinkish appearance owing to simultaneous contrast (Sect. 11.8) in the presence of the green surrounds. This pinkish color is, in turn, partially responsible for the appearance of complementary green in the afterimage seen against the gray background. The second process concerns the hardly noticeable pinkish afterimage of

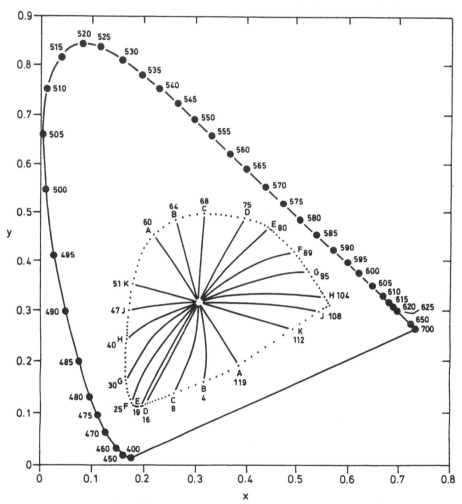

Fig. 11.5. Afterimage complementary hues. Pairs of lines of constant hue (*A* and *A*, *B* and *B*, etc.) indicate afterimage complementary pairs. The *black dots* represent the chromaticities of the colors of the original 120 samples. The numbers (e.g., 60 at *A*, 64 at *B*, ..., 119 at *A*, ...) identify individual samples. CIE ILL C. CIE 1931 (x, y) chromaticity diagram. (Based on [Ref. 5.21, Fig. 9])

the surrounds when viewed against the gray background. This pinkish afterimage is responsible for the induction of the remaining green into the square area by simultaneous contrast. This simple demonstration illustrates the complex nature of afterimages. Reconsideration of Plates XVIII–XIX in a more rigorous manner would require that such processes be taken into account. The colors that we perceive when we view a painting could be influenced significantly by the manner in which our gaze passes over it!

Of particular interest here is the investigation of afterimage complementary hues reported in 1955 by M.H. Wilson and R.W. Brocklebank [5.21]. They prepared 120 color samples of high purity (dyes on paper) which formed a complete hue circle with perceptually equal hue spacing (Fig. 11.5). They viewed the samples in daylight against a neutral gray surface and then perceived the afterimages while fixing their gaze on a neutral surface. While the afterimage was held clearly in view (for five to ten seconds), they measured the afterimage colors by comparing them with the colors produced by a spinning-disk technique.

In their tests, they established both afterimage complementary hues and additive complementary hues (Sect. 7.2). The results are represented by the two curves in Fig. 11.6. There is a tabulation of sample numbers

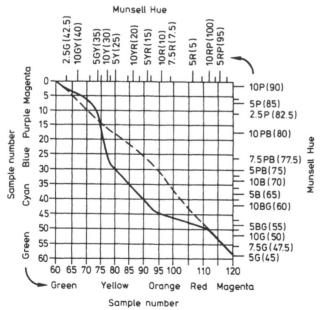

Fig. 11.6. Afterimage complementary pairs (- - -) and additive complementary pairs (——). The pairs are identified by sample numbers (uniform scales at the *left* and *bottom*) (Fig. 11.5) and approximately by Munsell number-letter Hue notation (and Munsell Hue number) (nonuniform scales at the *right* and *top*) (Fig. 8.11). An example of an afterimage complementary pair is sample 80 [5Y (25)] and sample 18 [9PB (79)], and an additive complementary pair is sample 80 [5Y (25)] and sample 30 [5PB (75)]. (Based on [Ref. 5.21, Fig. 5])

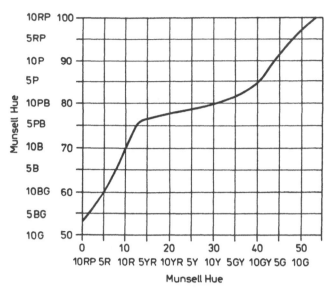

Fig. 11.7. Afterimage complementary pairs. The pairs are identified approximately by Munsell number-letter Hue notation (and by Munsell Hue number) (Fig. 8.11). The hues 10R (10) and 10B (70) are an example of an afterimage complementary pair. (Based on data in [Ref. 5.21, Fig. 7])

of half of the series of colors (0 to 60) in the color circle along the left vertical edge of the graph, and of the remaining numbers along the bottom horizontal edge. We can read, for example, that the additive complementary color of sample 30 (a blue) is a yellow (sample 80), and that its afterimage color is an orange (sample 95). Similarly, for sample 80 (yellow), the additive complementary is the blue (sample 30), and its afterimage complementary color is a purple (sample 18). The approximate Munsell Hues can be read on the scales along the top edge and the right vertical edge of the graph. A more convenient graph for determining directly the approximate Munsell Hues in afterimage complementary colors is presented in Fig. 11.7.

Wilson and Brocklebank repeated their tests days or weeks apart and found little variation of the results. Additional tests with samples of different purity and lightness and under varied conditions of illumination did not introduce any marked differences into the results.

In Fig. 11.5, lines of constant hue and varying purity determined in their experiments are shown connecting some of the pairs of samples that exhibited afterimage complementary hues (A–A, B–B, etc.). In three instances where the lines are straight (A–A, D–D, and K–K), the pairs represent both afterimage and additive complementary pairs. A perceived hue change occurs with change of purity along all *straight* radial lines (lines of additive color mixture) drawn from the central point (CIE ILL C), except for those that also connect afterimage pairs. Wilson and Brocklebank concluded, "Where additive complementary pairs are not the same as after-image pairs, the difference can be accounted for by the change of hue which takes place with additive mixing".

In the color circle of Wilson and Brocklebank, the hues are generally perceptually equally spaced except for minor adjustments that were made to bring afterimage complementary hues opposite each other. The spacing is visually more uniform than if additive complementary hues were opposites. This seems reasonable because a rather narrow range of yellow greens can neutralize ("cancel") chromatically (by additive color mixture) a compara-tively broad range of violets, and, similarly, a narrow range of blue greens can neutralize a wide range of reds [Note 11.3].

11.8 Simultaneous Contrast

Under normal viewing conditions, our eyes rove, rest, and move on again, a process that produces visual stimulation. When we fix our eyes on a colored patch in a painting for a sufficient period of time and then move them on to another patch of a different color, it is likely that the image of the first patch will be perceived upon the image of the second. But, even when our gaze moves slowly over a colored area, an afterimage that is much like a faded spot influences what is seen next [Refs. 11.8; 11.14, pp. 266, 269]. (From the foregoing, it is evident why the afterimage phenomenon is sometimes referred to by the term *successive contrast*.) The occurrence of afterimages is evidence of the variation of the response of the visual system with time.

Now let us consider a somewhat different situation. Let us fix our eyes on the square neutral gray patch at the center of the green area in Plate XX. We see that, with the green surround, the patch is not a neutral gray; it has a definite reddish cast. Indeed, we can verify that the color of the patch is a neutral gray in isolation if we look at it through a hole in a black reduction screen (one is provided at the back of the book). The red hue immediately perceived when our eyes fall on the square gray patch is an example of the phenomenon *simultaneous contrast*. Because it is difficult to control the motion of our eyes in normal viewing, the occurrence of what we take to be simultaneous contrast might include an afterimage contribution from the previous visual stimulation.

The occurrence of simultaneous contrast is made possible by the net-work structure of the retina. Although the receptor cones may be tiny enough to differentiate fine detail in an image, each is connected to nerve cells among which there are multiple interconnections [Refs. 2.3, p. 169; 2.6; 11.3]. In this way, light that falls on one point of the retina can affect signals from receptors in an area that surrounds that point. (For this reason, the phenomenon simultaneous contrast is sometimes referred to by the term *spatial contrast*.)

Jameson and Hurvich explain [3.6]: "... the visual tissue responds locally in a predictable way to focal stimulation of a given part of the retina, but ... this focal response activity is always subject to modification by ongoing activities in adjacent regions of visual tissue." Specifically they assume a *focal response* and an *induced response* from adjacent regions. "Thus localized 'greenness' activity in one part of the neural retina either suppresses 'greenness' activity in adjacent areas or accentuates 'redness' activity there. And the hypothesis is the same for blueness and yellowness, and for whiteness and blackness."

The influence of the color of a background on the color of a patch has been reported in books by Chevreul (1839) [1.6] and by Rood (1879) [1.7], which are classical works that continue to be of value in art and design. Rood has shown how a 10-member hue circle with afterimage complementary hue pairs diametrically opposite each other can be used to predict the influence. To illustrate his method, let us consider the influence of a red background (Munsell Hue 5R) on the perceived color of a test patch whose retinal image occupies the focal area. Four patch colors are taken as examples (Table 11.3).

Table 11.3. Simultaneous contrast. Influence of a red background (Munsell Hue 5R) on the perceived color of a patch

Patch	Patch color in isolation (aperture color)	Patch color influenced by red background
1	Yellow	Greenish yellow
2	Violet	Blue
3	Blue	Greenish blue
4	Blue green	Blue green

Starting with patch 1, imagine the position of the red on a hue circle and of the yellow relative to it. Rood's rule is that, when simultaneous contrast occurs, the perceived hue of the patch is displaced along the hue circle in the direction of the afterimage complementary of the color of the background, which in this case is blue green (Fig. 11.7). Hence, the hue of the patch is moved towards blue green, and the resulting perceived color is greenish yellow. This is reasonable if we consider that the hue induced by the red background, blue green, is "superimposed" on the focal yellow, accentuating the greenness activity and suppressing yellowness.

In the case of patch 2, we can consider that the green component of the induced response (blue green) suppresses some or all of the red component of the focal response (violet), leaving blue. Patch 3 has a greenish blue appearance, because the blue green induced response accentuates the blue focal response. Here no suppression takes place because no opponent components (red or yellow) are present. For patch 4 no change of hue is ex-

200

pected because, following the rule, no displacement is required to reach the complementary color; the induced response (blue green) adds to the focal response (blue green), which accentuates blue-greenness activity there.

Changes of saturation accompany the hue changes that result from color induction. Saturation increases when patch 4 is viewed against the red background. The induced blue green simply accentuates the focal blue green present, which increases the amount of blue green hue relative to the amount of white already present in the focal color. If, at the other extreme, we focus our eyes on the red background itself, the blue green response induced from the retinal area outside the focal area will suppress some of the focal red response, which results in perception of a less saturated red of the same hue. Following Rood, we find that the saturation of the perceived color of the patch depends upon its aperture color relative to the color of the background. If the aperture colors of a series of patches vary in hue from the background hue (red) through a sequence of hues along either half of the color circle toward the complementary (blue green) of the background red hue, the saturation of the color of a patch placed on the background shows a maximum decrease for the red patch and a maximum increase for the blue green patch. For some intermediate aperture color of a patch there must be a zero change of saturation, which suggests a hue boundary. For hues of patch colors between the background hue and the hue at the boundary, hue suppression occurs; for hues of patch colors between the hue at the boundary and the afterimage complementary hue, hue accentuation occurs.

These trends occur for colors of the patch and background that have nearly equal saturations. Let us now consider what can occur if the color of the patch is at reduced purity, for example a red patch on a background that has the same hue but higher purity. The perceived saturation of the patch is diminished because the induced blue green suppresses some of the focal red. It is possible that, in a second patch of still lower purity, the blue green may be just sufficient to suppress the focal red and leave a neutral gray appearance. In the third patch of purity lower than that of the second, the induced blue green will be more than enough to suppress the focal red; the patch will have a faint blue green appearance.

In a similar way, following Jameson and Hurvich, we can assume induced and focal response in terms of whiteness and blackness, which would result in *brightness contrast* [Ref. 2.3, p. 150]. Thus, strong stimulation in the background can inhibit activity in the focal receptors. So we find that a gray patch placed against a lighter gray or white background appears darker than when it is placed against a darker gray or black background. And, the color (chromatic) of a patch appears lighter and its saturation lower if it is placed against a background that is darker [11.15].

In the foregoing, attention is directed to the perceived color of a patch placed on a background. In this situation, the perceived color of the background is affected by the color of the patch, according to the same rule.

But these changes are not usually noticed, especially when the background is relatively large. Generally, if two different color patches of equal size are placed side by side, both will exhibit changes of appearance. If the patches are separated, the changes introduced diminish as the distance between them increases.

Jameson and Hurvich studied the occurrence of simultaneous contrast when the color of a square area was perceived in the presence of a second square of equal size but different color (with dark surrounds) [11.16]. They found that, when the squares were separated by a distance equal to their width, the contrast produced was equal to one-quarter of that measured when the squares touched along one side; when, in a square-grid arrangement, they were positioned so that only one corner touched, the contrast produced was equal to one-half. Contrast (or change) was measured in terms of chromatic response units (Sect. 11.2) under conditions of neutral adaptation [Refs. 2.3, p. 159; 11.16]. The observer's eye was focused on one square, one side of which subtended a visual angle of 3°.

Figure 11.8 summarizes the averaged results of this aspect of their investigation. It shows (for squares separated by a black gap equal to their width) how much the redness (or greenness) and yellowness (or blueness)

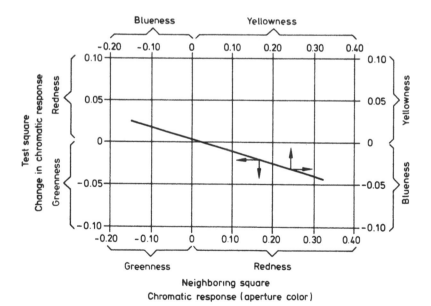

Fig. 11.8. Change of redness or greenness and yellowness or blueness from the aperture color of a test square on the fixation axis (Fig. 2.1) resulting from the introduction of a neighboring square of equal size separated by a black gap equal to their width. (An average color is assumed if the neighboring square is nonuniformly colored.) One side of the test square subtends a 3° visual angle. The indicated change is multiplied by 2 if the squares touch at a corner and by 4 if the squares abut along one side. (Based on [Ref. 11.16, Fig. 7])

of an inducing area affects the redness (or greenness) and yellowness (or blueness) of the focal area. Thus, for example, a redness of 0.16 chromatic units induces about 0.02 chromatic units of greenness in the focal area; a redness of 0.30 increases the greenness by 0.04, and so on. Thus, if the focal area is a red, or a reddish blue or a reddish yellow, its redness will be correspondingly diminished by 0.02 units (if the chromatic response of the inducing area is 0.16 units). Similarly, if the focal area is a green, blue, or greenish yellow, its greenness will be increased by 0.02 units.

This quantitative relation is of importance in the development of an opponent-colors theory that takes simultaneous contrast into account. It is relevant in art and design because it indicates in a quantitative way the magnitudes of the changes produced in simultaneous contrast and the influence of the separation of the inducing and focal areas on such changes.

11.9 Colored Shadows

Simultaneous contrast contributes in a major way to the richness of the gamut of colors that we experience in our surroundings. Josef Albers was keenly aware of this and exploited the phenomenon in his paintings [1.8–10]. Art students today are often introduced to the topic of simultaneous contrast by demonstrations with colored shadows. The shadows cast by diverse shapes of varying opacity illuminated by beams of colored light of different hue and purity can provide dramatic displays of varied colors [Refs. 2.3, p. 153; 7.43].

A striking demonstration can be given simply with a beam of white light and a beam of magenta light directed at an opaque object. Each beam casts a separate shadow on a white screen or wall in an otherwise darkened room. (Goethe proposed a similar demonstration [Ref. 1.1, Sect. 68, p. 30].) With such an arrangement, the unshaded portions of the screen appear pink, one shadow is magenta, and the other is green (Fig. 11.9). The shadow area that results from the white beam and the object (a dish, for example) is illuminated by the magenta beam. The shadow area that results from the magenta beam and the dish is illuminated by the white beam; it appears green. The green appearance is an example of the occurrence of simultaneous contrast: the green response induced by the surrounding pink surface is added to the white focal response produced by the shadow. We can prove that we are dealing with simultaneous contrast if we view the area of the green shadow in isolation, through a matt black paper tube. In this way, the pink area of the screen is excluded from view and the occurrence of the induced green response is prevented. Through the black tube we see only a portion of the screen illuminated by white light [Note 11.4].

The magenta color seen in the shadow may also show the influence of simultaneous contrast. The shadow area that results from the beam of

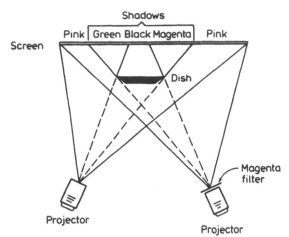

Fig. 11.9. Colored shadows. Diagram showing shadows on a screen in a darkened room produced by an opaque object (a dish) in the path of two light beams (one white, the other magenta). One shadow area is seen as green, another as magenta, and the central one (which receives little light) as black. The remainder of the screen illuminated by both beams has a pink appearance

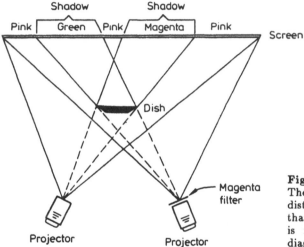

Fig. 11.10. Colored shadows. The projectors are shown at a distance from the screen such that a central black shadow is not formed. Compare this diagram with Fig. 11.9

white light and the dish is illuminated by the magenta beam, which produces a magenta focal response. This focal response is suppressed by the green response induced by the surrounding pink of the screen. The resulting magenta of the shadow is less saturated than the magenta seen through the black tube, which prevents an induced response.

When we look at a faithful color photograph made of the colored shadows on the screen, we see the induced green color. When, however, that shadow area of the photograph is viewed in isolation, as by use of a black reduction screen, the induced response is prevented, and green is not seen. This phenomenon occurs equally in pictures and in projected images.

Figure 11.9 shows also a black shadow in the region of the screen that does not receive light from either projection lamp (or, by reflection, from various parts of the room). The sizes of the black shadow and of the other shadows can be varied by changing the positions of the projectors and dish relative to the screen. In Fig. 11.10, the dish is shown sufficiently far from the screen so that the black shadow does not appear on it.

Here are several generalizations presented by Evans concerning the qualities of the colors that may be produced by induction in colored shadows [Ref. 2.5, p. 222]:

1. The effect of hue is optimized if the illuminance produced by the two light beams is about the same.
2. The saturation of the colors depends on the purity of the colors of the light beams. When there is a major difference of purity, as in the example of the white beam and the magenta beam, the higher purity (magenta) determines the saturation of the colors of the shadows (magenta and green).
3. If the two beams have the same dominant wavelength but their purities differ sufficiently, the color of the wall illuminated by both will be seen to have an intermediate saturation, the color of one shadow will be of the same hue but of higher saturation than that of the wall, and the color of the other shadow will be of the complementary hue.
4. The colors of the beams need not be distinctly different to produce complementary hues. For example, light beams from two ordinary lamps that differ only slightly in color temperature can produce bluish and yellowish shadows on a white wall.

In the early 1960s, much attention was attracted to two-color projection demonstrations by E.H. Land [Refs. 1.18, p. 371; 11.17,18]. He employed two projectors, one of which, for example, formed an image with a beam of red light (590–700 nm) and the other formed an image with a "white" beam (incandescent-lamp light); each image was of a black-and-white positive photographic slide. When the two images were accurately in register on a white screen, complex multicolored scenes (mostly still lifes) were seen. In some instances, the variety of colors produced was comparable to what could be obtained with three-color projections [Ref. 3.14, p. 444].

The two black-and-white positives were made from *color-separation photographs* of a scene. In the particular situation just described one color-separation photograph was made by photographing the scene with a red filter (585–700 nm) in front of the camera lens; the other color-separation photograph was made by use of a green filter (490–600 nm).

Judd has pointed out two important ways in which two-color projections are limited [11.8]. One is that many colors perceived to be different show up indistinguishably in two-color projection. For example, a dark gray (tree trunk), a yellow green (leaf), and a purple (flower petal) may all show up as a dark gray, unless memory color aids perception of the true color. The second is that many colors are not produced as perceived because of the restricted gamut possible with two-color projection. Judd has determined, for example, the following color discrepancies in a scene (Munsell Hue, Value, and Chroma designations): olive drab (Y 4/4) (military uniform), purple (RB 4/6) (iris petals), green (GY 5/8) (grass), and sky blue (B 9/4) would be reproduced under the best conditions of two-color projection (red beam and incandescent-lamp beam) as brown (YR 4/4), purplish black (RB 1/1), blue green (BG 5/8), and pale green (G 7/4), respectively. Although three-color projection, three-color photography, and three-color printing may also fail, for diverse reasons, two-color processes are doomed to a poorer performance because of a much more restricted gamut of colors.

Among the various phenomena that may be involved in two-color projections [Refs. 3.14, p. 444; 5.18, p. 110; 11.8,9,19–21], simultaneous contrast exerts a principal influence on what is seen. (Memory color can play a significant secondary role [11.21].) The occurrence of simultaneous contrast in the projections will be understood if we realize that colored shadows are cast on the screen by black images on the slides in the paths of the light beams [7.43]. The technique of two-color projection did not originate with Land. Evans in his earlier studies worked with it [Ref. 2.5,pp. 230, 233]. A report on two-color projection was published in 1897 [11.8], and it was employed by motion-picture producers as early as 1929 [11.20].

11.10 Edge Contrast

When simultaneous contrast occurs, the effect noted on a focal area seems to be uniform. For this reason, the effect is sometimes called *surface contrast*. On the other hand, if two juxtaposed areas of uniform colors having the same hue but slightly different luminance factor are viewed, it is found that, adjacent to the boundary between the two areas, there is a relative enhancement of lightness of the lighter area and a corresponding darkening of the adjoining darker area [Refs. 1.7, p. 268; 2.3, p. 164; 11.22, p. 194]. This can be demonstrated with a strip formed by a series of joined neutral gray squares, which form small uniform steps of decreasing luminance factor [Ref. 2.3, p. 164]. The effect is also demonstrated by either of the series of colored areas of approximately unchanging hue shown in Plate IX. Figure 11.11 shows the luminance-factor steps (a) and a corresponding hypothetical curve (c) that indicates schematically the variation of lightness perceived for one series. The uniformity of the printing of each area can be verified by viewing it in isolation by use of a black reduction screen. If a black line is drawn along the edge at which two areas join, the effect of edge contrast is lost (Fig. 11.11b).

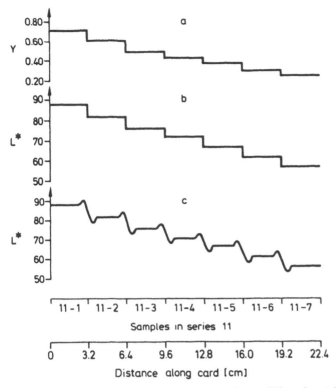

Fig. 11.11a–c. Edge contrast. Luminance factor (Y) and metric lightness (L^*) for part of a series of samples (series 11, Cascade lightness $L = 1$–7) from the *Munsell Limit Color Cascade* (Plate IX and Sect. 8.4). (**a**) Luminance factors of the samples (CIE ILL C). (**b**) Perceived lightness (metric lightness) of the samples without edge contrast (for example, each isolated or each separated from its neighbor by a black line) (Table 8.3). (**c**) Perceived lightness (metric lightness) of the samples with edge contrast (a hypothetical diagram expressing lightness enhancement and darkening on the two sides of the common border of abutting samples)

More than a century ago, the physicist Ernst Mach (1838–1916) called attention to edge effects and correctly deduced that they are caused physiologically by interactions among nerve cells in the retina [Ref. 2.3, p. 169]. The *edge contrast* phenomenon is also referred to by the terms *Mach-band effect, Mach contrast,* and *border contrast.*

Simultaneous contrast is explained by the hypothesis that the response to the stimulation of the focal area is modified by activity in neighboring retinal regions. As in the case of brightness contrast, strong stimulation of the neighboring retinal receptors can suppress the activity of the focal receptors to a low level. Edge contrast is also considered to arise because of these kinds of interplay among retinal cells. In Plate IX and Fig. 11.11c, for example, a peak of perceived lightness is explained by unequal influences from the left and right areas [Ref. 2.3, p. 174]. When the eye is focused on the lighter side of an edge, the response is suppressed less by the darker area on one side than by the lighter area on the other side. Hence the relatively high net response is made evident locally by the perceived increase of lightness. The opposite effect, a dip of lightness, is noted on the darker side where the net response is relatively low. There the response is suppressed more by the lighter area than by the darker area.

A striking example of heightened contrast is produced in a painting entitled "Arcturus" (1970) by Vasarely [Ref. 2.3, Plate 13-2]. The painting consists of four square panels, each of a different hue. For illustration here, one panel of diamond shape has been prepared in neutral grays (Plate XXI). The panel is made up of a series of equal bands or frames that present a sequence of steps from the center to the outside edge of the panel. The bands are colored uniformly in a progression of grays beginning with the lightest gray at the center. The "glowing" diagonals that extend across the panel were not printed. They are produced in the brain. An examination of a band in isolation (with a reduction screen) will reveal that there is no variation of perceived lightness.

In the foregoing instance, edge contrast is compounded at the intersections, which produces the glow. If the order of the grays is reversed, so that the outermost band is the lightest and the central diamond is the darkest, then no glow is produced. Instead, edge contrast is compounded to produce dark diagonal strips (Plate XXII). (In both illustrations, edge contrast is clearly apparent along the uniform bands.)

11.11 Assimilation (Reversed Contrast)

When we look at a page of printed words from a sufficient distance, we see a grayish blur because the detailed image is too small to be resolved by the mosaic of receptor cells in the retina. Artists often make use of this behavior of the eye when they employ cross-hatching in etchings and penned drawings to produce shading effects. The same applies to pointillistic paintings in which dots of selected colors are put on a surface to be perceived in spatial mixture (optical mixture) (Sect. 5.8).

When we approach a pointillistic painting closely enough to see individual dabs of paint, then the phenomenon of spatial mixture gives way to that of *assimilation*. The colors of the distinctly perceived dabs seem to shift towards each other. A demonstration of assimilation, an illustration made from three colored papers, is presented in Plate XXIII. The red strips on the yellow background appear yellowish; on the blue background, the red strips appear bluish. (The red rectangular area printed at the left with the same red ink is included to show the color without the influence of assimilation.) The phenomenon might be described by saying that the yellow and blue seem to have "spread" into the red. In this instance the hue of a red

Venetian blind (the red strips) changes as the man passes behind it, which gives the impression that the blind is translucent.

Assimilation is also known by the term *Bezold spreading effect* [Refs. 1.20, Fig. 3.12; 2.1, Plate XI; 2.3, p. 175, 2.6]. Assimilation is not simultaneous contrast; indeed it is sometimes called *reversed contrast*. In simultaneous contrast a red area surrounded by a yellow area would tend to look more bluish (not yellowish); surrounded by a blue area it would tend to look more yellowish (not bluish).

Why are two contrary effects displayed? Clearly it is related to the size of the images cast on the retina. Jameson and Hurvich propose that the explanation may be found in the different sizes of retinal *receptive fields*. A receptive field is a retinal region served by one interconnecting nerve cell which responds to stimuli imaged anywhere within the region [Ref. 2.3, p. 169]. Thus, if there are receptive fields that are sufficiently small compared with a retinal image, the resolution of the red strips (above) will be good and simultaneous contrast will be produced locally. If at the same time there are receptive fields that are so large that they cannot differentiate the detail of the strips, they will simply report an average response (for the red and yellow, or the red and blue). Thus the presence of large and small receptive fields would produce both the good resolution and average mixture observed. Assimilation involves the blending of hues and brightnesses, but with a maintenance of the perceived pattern. It is regarded as a kind of spatial averaging [Refs. 2.3, p. 176; 11.23].

At normal viewing distance, the detail of the retinal image of the grid pattern in Plate XXIII is adequate for assimilation to be experienced; the detail is too fine for simultaneous contrast to be dominant and not fine enough for pointillism. Let us now imagine ourselves viewing in its place another picture showing a pattern of bands of color of sufficient width to produce simultaneous contrast predominantly. Then as the viewing distance is increased, producing retinal images of finer detail, we note a transition from simultaneous contrast to assimilation. Finally a distance is reached beyond which the pattern is not distinguishable and we experience pointillism.

12. Appendix

12.1 Some Useful Addresses in the Field of Color

This list of addresses has been useful to me. A more extensive list may be found in [12.1].

A) Major Collections of Books on Color

1. Art and Architectural Library (The Faber Birren Collection on Color), Yale University, Box 1605A Yale Station, 180 York Street, New Haven CT 06520, USA [8.39–41]

2. Colour Reference Library, Royal College of Art, Kensington Gore, London SW7 2EU, England [8.42]

B) Color Research

3. JCRI: Japan Color Research Institute, 1–19 Nishiazabu 3 Chome, Minato-Ku, Tokyo 106, Japan

4. Munsell Color Science Laboratory, School of Photographic Arts and Sciences, RIT: Rochester Institute of Technology, 1 Lomb Memorial Drive, Rochester NY 14623, USA [12.2]

5. NRCC: Nationa' Research Council of Canada, Ottawa, Ontario, Canada K1A 0R6

6. Scandinavian Colour Institute, Riddargatan 17, P-O Box 14038, S–10440 Stockholm Sweden

C) Color Standards

7. ASTM: American Society for Testing and Materials, 1916 Race Street, Philadelphia, PA 19103, USA

8. BSI: British Standards Institution, 2 Park Street, London W1A 2BS, England

9. DIN: Deutsches Institut für Normung e.V., Beuth-Verlag GmbH, Burggraffenstrasse 4–10, Postfach 1107, D–1000 Berlin 30

10. Munsell Color, Macbeth Division of the Kollmorgen Corp., 2441 North Calvert Street, Baltimore MD 21218, USA [1.11,12]

11. NBS: National Bureau of Standards, Office of Standard Reference Materials, U.S. Department of Commerce, Washington DC 20234, USA

D) Associations

12. AATCC: American Association of Textile Chemists and Colorists, PO Box 12215, Research Triangle Park, NC 27709, USA

13. CIE: Bureau Central de la Commission Internationale de l'Eclairage, 52 Boulevard Malesherbs, 75008–Paris, France

14. FSCT: Federation of Societies for Coatings Technology, 1315 Walnut Street, Philadelphia PA 19107, USA

15. GATF: Graphic Arts Technical Foundation, 4615 Forbes Avenue, Pittsburgh PA 15213, USA

16. OSA: Optical Society of America, 1816 Jefferson Place NW, Washington DC 20036, USA

12.2 Conversion Table and Charts.
CIE(x, y, Y) Notation/Munsell Notation

A color specification CIE 1931 (x, y, Y) CIE ILL C can be converted to Munsell color notation (Hue, Value, Chroma – H V/C) (Sect. 8.4) by the use of Table 12.1 and one or two charts in the series presented in Figs. 12.1–9. The same table and charts may be used for the reversed conversion [Note 12.1].

Table 12.1 may be used to convert Munsell Value V to Y, which is the luminance factor, or luminous transmittance in the case of transparent materials. In the table, Y is expressed in percentages. For example at $V = 6.00$, $Y = 30.05\,\%$. In general practice, Y is often expressed as a fraction, $Y = 0.3005$.

The Munsell-CIE charts in Figs. 12.1–9 provide constant-Value definitions of Munsell Hue and Munsell Chroma in terms of the CIE 1931 (x, y) chromaticity diagram. In each chart, only part of the chromaticity diagram is shown; that part includes most, if not all, of the domain of practical interest.

All the charts show lines (slightly curved, for the most part) that radiate from the point (0.3101, 0.3162), which represents the chromaticity of CIE ILL C. (The charts can be used only for Munsell colors that are observed in daylight.) The radial lines are lines of constant Munsell Hue. They correspond to the Hue radii in the Munsell Hue circle (Fig. 8.9). (But note that, in the charts, the sequence in the Munsell Hue circle is presented in a clockwise manner; in Figs. 12.1–9 the Hues progress in a counter-clockwise manner.) The Hue lines are shown at intervals of 2.5 Hue units; they are identified by the Munsell number-letter notation. At five-unit intervals they are also identified by Munsell Hue numbers, which are indicated within parentheses. Thus, in a Hue range of 10 units, we read, for example, the sequence 10R(10), 2.5YR, 5YR(15), 7.5YR, 10YR(20). It should be remembered that the end of the R series 10R(10) coincides with the beginning of the YR series 0YR(10); the end of the YR series 10YR(20) coincides with the beginning of the Y series 0Y(20), and so on.

The ovals are lines of constant Munsell Chroma. They are shown at intervals of two Chroma units (2, 4, 6, ...). A supplementary series of charts for a low Chroma range (0.5, 1, 1.5, 2, 2.5, ...) and for Values $V = 5$ to 9 may be found in [8.19,19a]; the entire series may be found in large format in [8.44] [Note 12.1].

The charts do not show Chroma and Hue lines beyond the MacAdam limits. In Fig. 12.9 (Value 9), we can see a large portion of the MacAdam limits included within the edges of the chart. Also evident is a portion of the spectrum locus in the upper right-hand corner and a portion of the purple line in the lower right-hand corner.

Two examples of the conversion from CIE(x, y, Y) to Munsell notation:

1) Find the equivalent Munsell notation for CIE(x, y, Y) = CIE $(0.250, 0.350, 0.300)$.

a) The first step is to find the Munsell Value. Given Y (the luminance factor, or luminous transmittance if the object is transparent), we can find in Table 12.1 the corresponding Value V. In the CIE specification, Y is given as a fraction, 0.300; to use the table it should be converted to a percentage by multiplying by 100, hence 30.0%. Because the table shows $V = 6.00$ at $Y = 30.05\%$, we accept $V = 6.00$ as the conversion. [If Y were 0.311 (31.1%), we would note that $V = 6.08$ (for $Y = 30.99\%$) and $V = 6.09$ (for $Y = 31.11\%$) and would accept $V = 6.09$ for the conversion.]

b) The second step is to determine the Munsell Hue H and the Munsell Chroma C. Figure 12.6, which applies specifically to Munsell Value $V = 6$, is employed for this purpose. The chromaticity point (0.250, 0.350) is plotted on the chart at $x = 0.250$, $y = 0.350$. Munsell H and C are determined by the position of the point relative to the two radial Hue lines and the two Chroma arcs that enclose it. The two enclosing Hue lines are labeled 10G [which is equivalent to 0BG (Fig. 8.9)] and 2.5BG. The point is about 3/4 of the distance from 0BG to 2.5BG, so we may judge the Hue to be 3/4 of 2.5, or about 1.8BG [or, Hue number: 51.8 (Fig. 8.11)].

c) The two enclosing Chroma arcs are at Chroma 6 and 8. The position of the point is about 1/5 of the radial distance (2 units) from $C = 6$ to $C = 8$, 2/5 of a unit more than 6. Hence the Chroma of the color is 6 plus 2/5, or 6.4.

d) The Munsell notation for the color is 1.8BG 6.0/6.4.

In the above example, Value V was found to be essentially 6.00, and hence it was possible to use Fig. 12.6 ($V = 6$) directly to determine Munsell Hue and Chroma. If the Value were slightly different, say up to 6.05 or down to 5.95, we could still use Fig. 12.6 for the determination. But, generally speaking, it is necessary to use two charts. Thus, in the foregoing example, if $V = 6.09$, it would be necessary to use the charts for $V = 7$ and $V = 6$ (Figs. 12.7 and 6). In the following example in which V is found to be 5.40, the charts for $V = 6$ and $V = 5$ (Figs. 12.6 and 5) are required.

2) Find the equivalent Munsell notation for CIE(x, y, Y) =CIE $(0.400, 0.400, 0.236)$.

a) From Table 12.1 we find $V = 5.40$ (corresponding to $Y = 0.2357$ or 23.57%).

b) From chart $V = 6$ (Fig. 12.6), we learn that the Hue is 2.5Y. In this case, the estimation of Hue is simplified because the point happens to fall on the radial Hue line 2.5Y. Chroma is found to be 4.8. See 1(b), (c).

c) From chart $V = 5$ (Fig. 12.5), we find the Hue to be 3Y and Chroma, 4.1. See 1(b), (c).

d) Because the Value $V = 5.40$ is 0.40 of the distance between $V = 5$ and $V = 6$, we take the Hue at 5.40 to be 0.40 of the distance from 3Y down to 2.5Y – that is, 0.40 times 0.5, or 0.2. Hence, stepping down 0.2 Hue units from 3Y leads to a Hue of 2.8Y (Hue number 22.8).

e) Similarly, in passing from $C = 4.1$ to $C = 4.8$, the change of Chroma is 0.7 units, which corresponds to the change from $V = 5$ to $V = 6$. At $V = 5.40$ the Chroma is 0.40 of the distance from 4.1 to 4.8 – hence, 0.40 times 0.7, or 0.28 Chroma units. Stepping up 0.28 Chroma units from 4.1 gives Chroma 4.38, or simply 4.4.

f) The Munsell notation for the color is 2.8Y 5.4/4.4.

Thus, beginning with a $CIE(x, y, Y)$ color specification, we can obtain the corresponding Munsell notation. Having the Munsell notation, we can proceed one step further to find the ISCC-NBS color name (Sect. 12.3 and Tables 10.1 and 12.2).

In instances where it is necessary to convert Munsell notation to CIE (x, y, Y) notation, the procedure is reversed. Table 12.1 is used to convert V directly to Y (as a percentage). If $V = 5.40$, a point that corresponds to H and C is plotted on the charts for $V = 5$ (Fig. 12.5) and $V = 6$ (Fig. 12.6), and the pairs of values for x and y are read off. Thus, if H is 2Y and C is 3.5, (x, y) for $V = 5$ is (0.386, 0.383) and for $V = 6$, (0.375, 0.376). To find (x, y) for 5.40, note that x is 0.40 of the distance from 0.386 to 0.375, and y is 0.40 of the distance from 0.383 to 0.376. Thus the $CIE(x, y, Y)$ notation is CIE 1931 (0.382, 0.380, 0.236) CIE ILL C.

Table 12.1. Conversion table. Munsell Value V; luminance factor Y (reflectance) or luminous transmittance Y.[Ref. 8.20, Table II]

V	Y [%]	V	Y [%]	V	Y [%]	V	Y [%]	V	Y [%]
10.00	102.56								
9.99	102.30	9.54	90.97	9.09	80.62	8.64	71.17	8.19	62.52
8	102.04	3	90.73	8	80.40	3	70.97	8	62.34
7	101.78	2	90.49	7	80.18	2	70.77	7	62.16
6	101.52	1	90.25	6	79.97	1	70.57	6	61.98
5	101.25	0	90.01	5	79.75	0	70.37	5	61.79
9.94	100.99	9.49	89.77	9.04	79.53	8.59	70.17	8.14	61.61
3	100.73	8	89.53	3	79.31	8	69.97	3	61.43
2	100.47	7	89.30	2	79.10	7	69.78	2	61.25
1	100.21	6	89.06	1	78.88	6	69.58	1	61.07
0	99.95	5	88.82	0	78.66	5	69.38	0	60.88
9.89	99.69	9.44	88.59	8.99	78.45	8.54	69.18	8.09	60.70
8	99.44	3	88.35	8	78.23	3	68.99	8	60.52
7	99.18	2	88.12	7	78.02	2	68.79	7	60.35
6	98.92	1	87.88	6	77.80	1	68.59	6	60.17
5	98.66	0	87.65	5	77.59	0	68.40	5	59.99
9.84	98.41	9.39	87.41	8.94	77.38	8.49	68.20	8.04	59.81
3	98.15	8	87.18	3	77.16	8	68.01	3	59.63
2	97.90	7	86.95	2	76.95	7	67.81	2	59.45
1	97.64	6	86.72	1	76.74	6	67.62	1	59.28
0	97.39	5	86.48	0	76.53	5	67.43	0	59.10
9.79	97.14	9.34	86.25	8.89	76.32	8.44	67.23	7.99	58.92
8	96.88	3	86.02	8	76.11	3	67.04	8	58.74
7	96.63	2	85.79	7	75.90	2	66.85	7	58.57
6	96.38	1	85.56	6	75.69	1	66.66	6	58.39
5	96.13	0	85.33	5	75.48	0	66.46	5	58.22
9.74	95.88	9.29	85.10	8.84	75.27	8.39	66.27	7.94	58.04
3	95.63	8	84.88	3	75.06	8	66.08	3	57.87
2	95.38	7	84.65	2	74.85	7	65.89	2	57.69
1	95.13	6	84.42	1	74.64	6	65.70	1	57.52
0	94.88	5	84.19	0	74.44	5	65.51	0	57.35
9.69	94.63	9.24	83.97	8.79	74.23	8.34	65.32	7.89	57.17
8	94.38	3	83.74	8	74.02	3	65.13	8	57.00
7	94.14	2	83.52	7	73.82	2	64.94	7	56.83
6	93.89	1	83.29	6	73.61	1	64.76	6	56.66
5	93.64	0	83.07	5	73.40	0	64.57	5	56.48
9.64	93.40	9.19	82.84	8.74	73.20	8.29	64.38	7.84	56.31
3	93.15	8	82.62	3	72.99	8	64.19	3	56.14
2	92.91	7	82.39	2	72.79	7	64.01	2	55.97
1	92.66	6	82.17	1	72.59	6	63.82	1	55.80
0	92.42	5	81.95	0	72.38	5	63.63	0	55.63
9.59	92.18	9.14	81.73	8.69	72.18	8.24	63.45	7.79	55.46
8	91.93	3	81.50	8	71.98	3	63.26	8	55.29
7	91.69	2	81.28	7	71.78	2	63.08	7	55.12
6	91.45	1	81.06	6	71.57	1	62.89	6	54.95
5	91.21	0	80.84	5	71.37	0	62.71	5	54.78

Table 12.1 (continued)

V	Y [%]	V	Y [%]	V	Y [%]	V	Y [%]	V	Y [%]
7.74	54.62	7.29	47.38	6.84	40.79	6.39	34.79	5.94	29.36
3	54.45	8	47.23	3	40.65	8	34.66	3	29.25
2	54.28	7	47.08	2	40.51	7	34.54	2	29.13
1	54.11	6	46.92	1	40.37	6	34.41	1	29.02
0	53.94	5	46.77	0	40.23	5	34.28	0	28.90
7.69	53.78	7.24	46.62	6.79	40.09	6.34	34.16	5.89	28.79
8	53.61	3	46.47	8	39.95	3	34.03	8	28.68
7	53.45	2	46.32	7	39.82	2	33.91	7	28.57
6	53.28	1	46.17	6	39.68	1	33.78	6	28.45
5	53.12	0	46.02	5	39.54	0	33.66	5	28.34
7.64	52.95	7.19	45.87	6.74	39.40	6.29	33.54	5.84	28.23
3	52.79	8	45.72	3	39.27	8	33.41	3	28.12
2	52.62	7	45.57	2	39.13	7	33.29	2	28.01
1	52.46	6	45.42	1	39.00	6	33.16	1	27.90
0	52.30	5	45.27	0	38.86	5	33.04	0	27.78
7.59	52.13	7.14	45.12	6.69	38.72	6.24	32.92	5.79	27.67
8	51.97	3	44.97	8	38.59	3	32.80	8	27.56
7	51.81	2	44.82	7	38.45	2	32.67	7	27.45
6	51.64	1	44.67	6	38.32	1	32.55	6	27.34
5	51.48	0	44.52	5	38.18	0	32.43	5	27.23
7.54	51.32	7.09	44.38	6.64	38.05	6.19	32.31	5.74	27.12
3	51.16	8	44.23	3	37.92	8	32.19	3	27.02
2	51.00	7	44.08	2	37.78	7	32.07	2	26.91
1	50.84	6	43.94	1	37.65	6	31.95	1	26.80
0	50.68	5	43.79	0	37.52	5	31.83	0	26.69
7.49	50.52	7.04	43.64	6.59	37.38	6.14	31.71	5.69	26.58
8	50.36	3	43.50	8	37.25	3	31.59	8	26.48
7	50.20	2	43.35	7	37.12	2	31.47	7	26.37
6	50.04	1	43.21	6	36.99	1	31.35	6	26.26
5	49.88	0	43.06	5	36.86	0	31.23	5	26.15
7.44	49.72	6.99	42.92	6.54	36.72	6.09	31.11	5.64	26.05
3	49.56	8	42.77	3	36.59	8	30.99	3	25.94
2	49.41	7	42.63	2	36.46	7	30.87	2	25.84
1	49.25	6	42.49	1	36.33	6	30.75	1	25.73
0	49.09	5	42.34	0	36.20	5	30.64	0	25.62
7.39	48.93	6.94	42.20	6.49	36.07	6.04	30.52	5.59	25.52
8	48.78	3	42.06	8	35.94	3	30.40	8	25.41
7	48.62	2	41.92	7	35.81	2	30.28	7	25.31
6	48.47	1	41.77	6	35.68	1	30.17	6	25.20
5	48.31	0	41.63	5	35.56	0	30.05	5	25.10
7.34	48.16	6.89	41.49	6.44	35.43	5.99	29.94	5.54	25.00
3	48.00	8	41.35	3	35.30	8	29.82	3	24.89
2	47.85	7	41.21	2	35.17	7	29.71	2	24.79
1	47.69	6	41.07	1	35.03	6	29.59	1	24.69
0	47.54	5	40.93	0	34.92	5	29.48	0	24.58

Table 12.1 (continued)

V	Y [%]	V	Y [%]	V	Y [%]	V	Y [%]	V	Y [%]
5.49	24.48	5.04	20.13	4.59	16.29	4.14	12.93	3.69	10.075
8	24.38	3	20.04	8	16.21	3	12.86	8	10.017
7	24.28	2	19.95	7	16.13	2	12.80	7	9.959
6	24.17	1	19.86	6	16.05	1	12.73	6	9.901
5	24.07	0	19.77	5	15.97	0	12.66	5	9.843
5.44	23.97	4.99	19.68	4.54	15.89	4.09	12.59	3.64	9.785
3	23.87	8	19.59	3	15.81	8	12.52	3	9.728
2	23.77	7	19.50	2	15.74	7	12.46	2	9.671
1	23.67	6	19.41	1	15.66	6	12.39	1	9.614
0	23.57	5	19.32	0	15.57	5	12.32	0	9.557
5.39	23.47	4.94	19.23	4.49	15.49	4.04	12.26	3.59	9.501
8	23.37	3	19.14	8	15.42	3	12.19	8	9.445
7	23.27	2	19.06	7	15.34	2	12.12	7	9.389
6	23.17	1	18.97	6	15.26	1	12.06	6	9.333
5	23.07	0	18.88	5	15.18	0	12.00	5	9.277
5.34	22.97	4.89	18.79	4.44	15.11	3.99	11.935	3.54	9.222
3	22.87	8	18.70	3	15.03	8	11.870	3	9.167
2	22.78	7	18.62	2	14.96	7	11.805	2	9.112
1	22.68	6	18.53	1	14.88	6	11.740	1	9.058
0	22.58	5	18.44	0	14.81	5	11.675	0	9.003
5.29	22.48	4.84	18.36	4.39	14.73	3.94	11.611	3.49	8.949
8	22.38	3	18.27	8	14.66	3	11.547	8	8.895
7	22.29	2	18.19	7	14.58	2	11.483	7	8.841
6	22.19	1	18.10	6	14.51	1	11.419	6	8.787
5	22.09	0	18.02	5	14.43	0	11.356	5	8.734
5.24	22.00	4.79	17.93	4.34	14.36	3.89	11.292	3.44	8.681
3	21.90	8	17.85	3	14.28	8	11.229	3	8.628
2	21.81	7	17.76	2	14.21	7	11.167	2	8.575
1	21.71	6	17.68	1	14.14	6	11.104	1	8.523
0	21.62	5	17.60	0	14.07	5	11.042	0	8.471
5.19	21.52	4.74	17.51	4.29	13.99	3.84	10.980	3.39	8.419
8	21.43	3	17.43	8	13.92	3	10.918	8	8.367
7	21.33	2	17.34	7	13.85	2	10.856	7	8.316
6	21.24	1	17.26	6	13.78	1	10.795	6	8.264
5	21.14	0	17.18	5	13.70	0	10.734	5	8.213
5.14	21.05	4.69	17.10	4.24	13.63	3.79	10.673	3.34	8.162
3	20.96	8	17.02	3	13.56	8	10.612	3	8.111
2	20.86	7	16.93	2	13.49	7	10.551	2	8.060
1	20.77	6	16.85	1	13.42	6	10.491	1	8.010
0	20.68	5	16.77	0	13.35	5	10.431	0	7.960
5.09	20.59	4.64	16.69	4.19	13.28	3.74	10.371	3.29	7.910
8	20.49	3	16.61	8	13.21	3	10.311	8	7.860
7	20.40	2	16.53	7	13.14	2	10.252	7	7.811
6	20.31	1	16.45	6	13.07	1	10.193	6	7.762
5	20.22	0	16.37	5	13.00	0	10.134	5	7.713

Table 12.1 (continued)

V	Y [%]	V	Y [%]	V	Y [%]	V	Y [%]	V	Y [%]
3.24	7.664	2.79	5.680	2.34	4.092	1.89	2.853	1.44	1.910
3	7.615	8	5.641	3	4.060	8	2.829	3	1.892
2	7.567	7	5.602	2	4.029	7	2.805	2	1.874
1	7.519	6	5.563	1	3.998	6	2.781	1	1.856
0	7.471	5	5.524	0	3.968	5	2.758	0	1.838
3.19	7.423	2.74	5.485	2.29	3.938	1.84	2.735	1.39	1.821
8	7.375	3	5.447	8	3.907	3	2.712	8	1.803
7	7.328	2	5.408	7	3.877	2	2.688	7	1.786
6	7.281	1	5.370	6	3.847	1	2.665	6	1.769
5	7.234	0	5.332	5	3.817	0	2.642	5	1.752
3.14	7.187	2.69	5.295	2.24	3.787	1.79	2.620	1.34	1.735
3	7.140	8	5.257	3	3.758	8	2.598	3	1.718
2	7.094	7	5.220	2	3.729	7	2.575	2	1.701
1	7.048	6	5.183	1	3.700	6	2.553	1	1.684
0	7.002	5	5.146	0	3.671	5	2.531	0	1.667
3.09	6.956	2.64	5.109	2.19	3.642	1.74	2.509	1.29	1.650
8	6.911	3	5.072	8	3.613	3	2.487	8	1.634
7	6.866	2	5.036	7	3.585	2	2.465	7	1.618
6	6.821	1	5.000	6	3.557	1	2.443	6	1.601
5	6.776	0	4.964	5	3.529	0	2.422	5	1.585
3.04	6.731	2.59	4.928	2.14	3.501	1.69	2.401	1.24	1.569
3	6.687	8	4.892	3	3.473	8	2.380	3	1.553
2	6.643	7	4.857	2	3.445	7	2.359	2	1.537
1	6.599	6	4.822	1	3.418	6	2.338	1	1.521
0	6.555	5	4.787	0	3.391	5	2.317	0	1.506
2.99	6.511	2.54	4.752	2.09	3.364	1.64	2.296	1.19	1.490
8	6.468	3	4.717	8	3.337	3	2.276	8	1.475
7	6.425	2	4.682	7	3.310	2	2.256	7	1.459
6	6.382	1	4.648	6	3.283	1	2.236	6	1.444
5	6.339	0	4.614	5	3.256	0	2.216	5	1.429
2.94	6.296	2.49	4.580	2.04	3.230	1.59	2.196	1.14	1.413
3	6.254	8	4.546	3	3.204	8	2.176	3	1.398
2	6.212	7	4.512	2	3.178	7	2.156	2	1.383
1	6.170	6	4.479	1	3.152	6	2.136	1	1.368
0	6.128	5	4.446	0	3.126	5	2.116	0	1.354
2.89	6.086	2.44	4.413	1.99	3.100	1.54	2.097	1.09	1.339
8	6.045	3	4.380	8	3.075	3	2.078	8	1.324
7	6.003	2	4.347	7	3.050	2	2.059	7	1.310
6	5.962	1	4.314	6	3.025	1	2.040	6	1.295
5	5.921	0	4.282	5	3.000	0	2.021	5	1.281
2.84	5.881	2.39	4.250	1.94	2.975	1.49	2.002	1.04	1.267
3	5.841	8	4.218	3	2.950	8	1.983	3	1.253
2	5.800	7	4.186	2	2.925	7	1.965	2	1.238
1	5.760	6	4.154	1	2.901	6	1.947	1	1.224
0	5.720	5	4.123	0	2.877	5	1.929	0	1.210

Table 12.1 (continued)

V	Y [%]	V	Y [%]	V	Y [%]	V	Y [%]	V	Y [%]
0.99	1.196	0.79	0.931	0.59	0.687	0.39	0.455	0.19	0.225
8	1.182	8	0.918	8	0.675	8	0.444	8	0.214
7	1.168	7	0.906	7	0.663	7	0.432	7	0.202
6	1.154	6	0.893	6	0.651	6	0.421	6	0.191
5	1.141	5	0.881	5	0.640	5	0.409	5	0.179
0.94	1.128	0.74	0.868	0.54	0.628	0.34	0.398	0.14	0.167
3	1.114	3	0.856	3	0.617	3	0.386	3	0.155
2	1.101	2	0.844	2	0.605	2	0.375	2	0.143
1	1.087	1	0.832	1	0.593	1	0.363	1	0.131
0	1.074	0	0.819	0	0.581	0	0.352	0	0.120
0.89	1.060	0.69	0.807	0.49	0.570	0.29	0.341	0.09	0.108
8	1.047	8	0.795	8	0.559	8	0.329	8	0.096
7	1.034	7	0.783	7	0.547	7	0.318	7	0.084
6	1.021	6	0.771	6	0.535	6	0.306	6	0.073
5	1.008	5	0.759	5	0.524	5	0.295	5	0.061
0.84	0.995	0.64	0.747	0.44	0.513	0.24	0.283	0.04	0.049
3	0.982	3	0.735	3	0.501	3	0.272	3	0.036
2	0.969	2	0.723	2	0.489	2	0.260	2	0.024
1	0.956	1	0.711	1	0.478	1	0.248	1	0.012
0	0.943	0	0.699	0	0.467	0	0.237	0	0.000

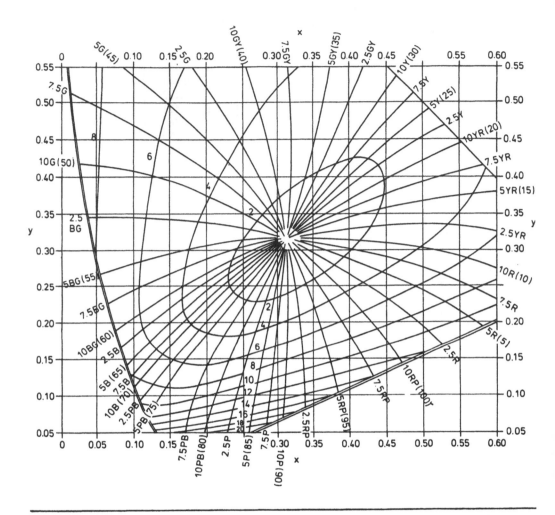

Fig. 12.1

Figs. 12.1-9. Munsell-CIE charts. Lines of constant Munsell Hue and constant Munsell Chroma on the CIE 1931 (x, y) chromaticity diagram at Munsell Values 1-9, respectively. Hue is indicated near the end of each radiating line. (Hue number is given within parentheses.) Chroma is indicated on the ovals in steps of two. (Based on charts prepared by Dorothy Nickerson [Note 12.1])

220

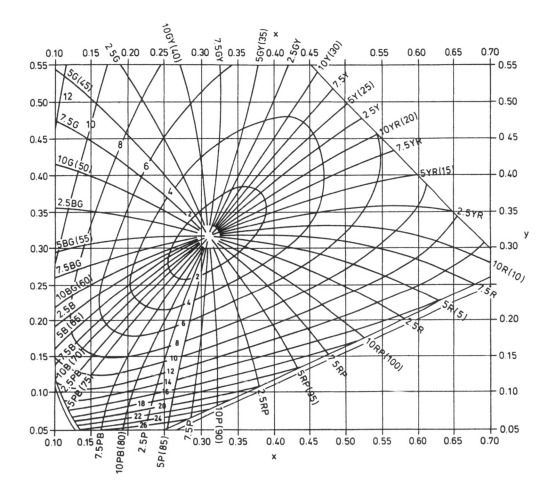

Fig. 12.2

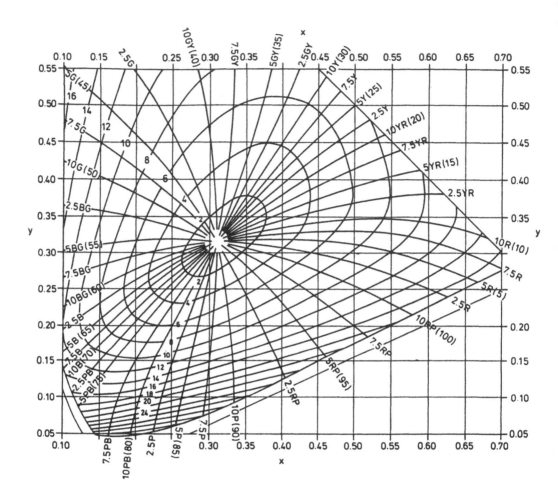

Fig. 12.3

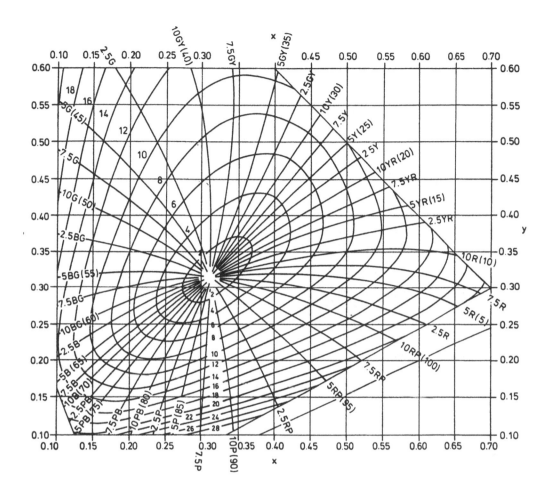

Fig. 12.4

223

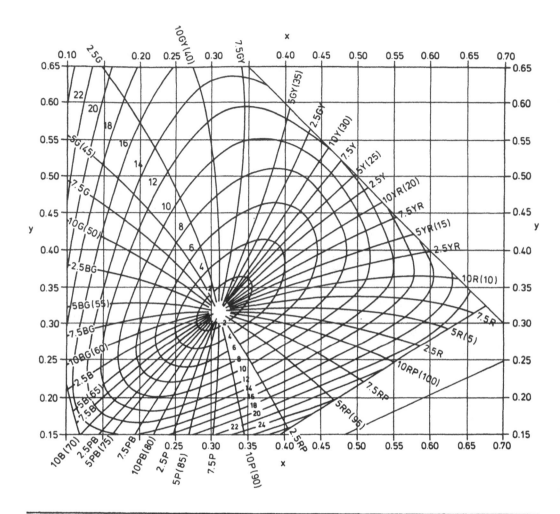

Fig. 12.5

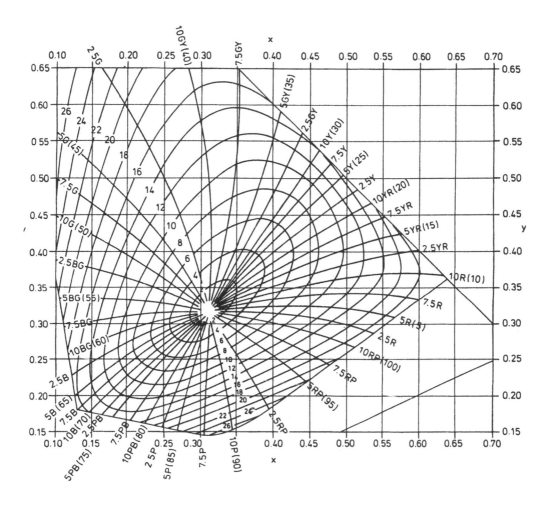

Fig. 12.6

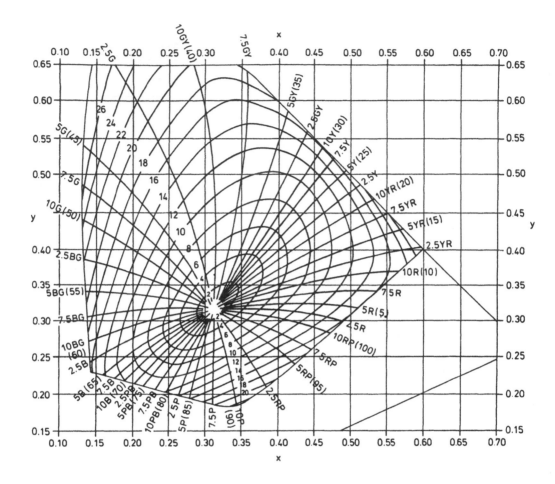

Fig. 12.7

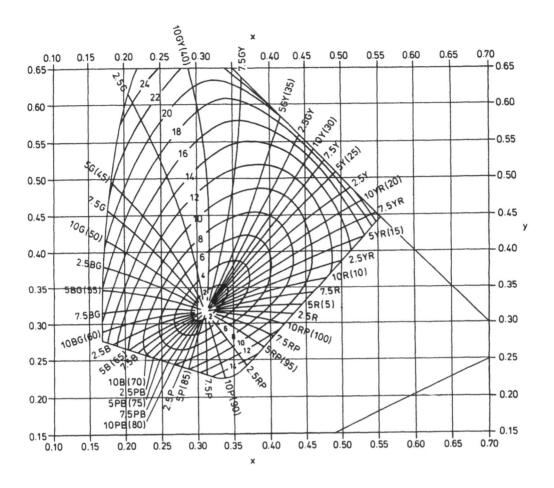

Fig. 12.8

227

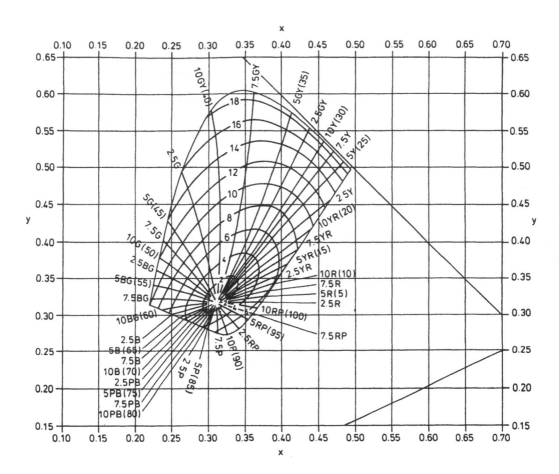

Fig. 12.9

12.3 Procedure for Determining ISCC-NBS Color Names

The procedure for determining the ISCC-NBS color name for the color of an object requires a specification of the color in terms of Munsell Hue, Value, and Chroma. [A specification in CIE(x, y, Y) notation may be converted to Munsell Hue, Value, and Chroma by using the method described in Sect. 12.2.] Given the Munsell specification, we can find the ISCC-NBS centroid number of the color with the use of Table 12.2. Having found the centroid number, we can find the corresponding ISCC-NBS color name in Table 10.1. This section concerns primarily the use of Table 12.2.

Let us first consider the arrangement of Table 12.2. The first column (at the left) presents the Value range for each of the color-name blocks (Sect. 10.1). The second set of columns gives the particular numerical values of Chroma (0.0, 0.5, 0.7, 1, 1.2, ..., 6, 7, 8, ..., 13, 14, 15, 40) that are used to designate the Chroma range of each of the blocks. The third set of columns gives the particular Munsell Hues (for example, 9PB, 3P, 9P, ..., 1R) used to designate the Hue range of a block. To illustrate this arrangement, let us consider the color-name block applying to the ISCC-NBS color name "light purple" (centroid number 222) in Fig. 10.1. The Value range ($V = 5.5-7.5$) represents the height of the color-name block, the Chroma range ($C = 5-9$) gives its radial ("length") dimension, and the Hue range ($H = 3P-9P$) describes its angular ("width") dimension. These three ranges, which describe the color-name block with which centroid number 222 is associated, are indicated in Table 12.2. The centroid numbers are given in the right-hand column of the table. In principle, the procedure (described below) involves finding the color block (and, thereby, the centroid number) in which the color (specified in Munsell notation) is located. Hence, if the Munsell specification is 4P 6/8, the Munsell Hue, Value, and Chroma will be found to fall within the ranges given above, and the point in Munsell color space will fall in this color-name block (centroid number 222).

Some centroid numbers appear two or three times in the last column. The reason is that, in those cases, the blocks have more than six faces, for example the block for "pale purple" (centroid number 227) in Fig. 10.1. To simplify Table 12.2, these "odd" blocks have been cut into two or three six-faced forms, each of which can be described by one set of three ranges of Value, Chroma, and Hue. Other simplifications employed are the use of the upper Value limit 10, the lower Value limit 0, and the arbitrary upper Chroma limit 40 [Ref. 8.43, Table I] wherever a definite Value or Chroma boundary for a color-name block at the limit of a color gamut is not indicated in the ISCC-NBS color-name charts.

To illustrate the procedure for using Table 12.2, let us consider a color for which the Munsell designation is 5R 7/4. There are several steps:

1) Consult the pages of Table 12.2 on which the Hue in question is included. Hue 5R is included within the ranges shown on pp. 231 and 232.

2) Note the tabulated ranges of Value that include the Value of interest. For Value 7, the following ranges must be considered on p. 231: 6.5–10, 6.5–8.5, 6.5–8.0, and 6.5–7.5. On p. 232 there is one additional range to consider: 4.5–10.

3) In parts of the table where the ranges of Value are considered, simultaneously trace downward between the two consecutive columns of Chroma (i.e., Chromas 3 and 5) that bracket the Chroma of interest (Chroma 4) and between the two columns of Hue (i.e., Hues 4R and 6R) that bracket the Hue of interest (Hue 5R). On the horizontal line on which *both* Chroma 4 and Hue 5R are bracketed by 0's, the last column gives the centroid number sought: 5. The color name is then found by consulting Table 10.1. For centroid number 5, the color is shown to be "moderate pink".

Colors that are only slightly different are often found to be in the same color-name block in Munsell color space; hence, they share the same ISCC-NBS color name and centroid number. For example, color samples I-A and I-B in Plate V have the same color name (Table 7.4).

Sometimes, in Munsell color space, the location of a color is on a boundary between the color-name blocks. In this case, two ISCC-NBS color names apply. Cases may be found in which as many as eight color names apply, as for the color whose Munsell notation is 7Y 8/8. Here, the location of the color in Munsell color space is at a point where the corners of eight color-name blocks touch.

To avoid the ambiguity of multiple color names, it seems generally preferable to alter very slightly, but in a consistent way, all quantities in a Munsell notation that lead to more than one color name. I suggest that Value be increased by 0.1 if its amount is given by one of the following numbers: 1.5, 2.0, 2.5, 3.0, 3.5, 4.5, 5.5, 6.5, 7.5, 8.0, and 8.5. Thus if 7.5 is given for the Value, then it would be taken as 7.6 for the purpose of determining a single color name. Similarly, if a Chroma or a Hue is numerically identical to one appearing at the heads of columns in Table 12.2, then it would be increased by 0.1 to avoid finding more than one color name. Thus Chroma 9 is raised to 9.1 for sample 6-B in Table 7.3, leading to the color name "strong blue" (178). Similarly Hue 5PB is changed to 5.1PB, and the color name for sample 6-C is found to be "light purplish blue" (199).

Table 12.2. Key for converting a Munsell notation (Value, Chroma, Hue) into an ISCC-NBS centroid number (No.). This table presents in tabular form the information presented in graphical form in 31 *color-name charts* [Ref. 12.3, pp. 16–31]. The ISCC-NBS color names are found through the centroid numbers in Table 10.1 of this book

```
VALUE                     CHROMA                                    HUE              NO.
        0 0 0 1 1 1 2 2 3 5 6 7 8 9 1 1 1 1 1 4  9 1 4 6 7 8 9 1 2 3 5 7 8
        . . . . . . . .               0 1 3 4 5 0  R R R R R R R Y Y Y Y Y Y
        0 5 7 0 2 5 0 5                            P           R R R R R R

8.5-10. 0 0 . . . . . . . . . . . . . . . . . .  0 0 0 0 0 0 0 0 0 0 0 . 263
8.5-10. 0 0 0 . . . . . . . . . . . . . . . . .  . . . . . . . . . . 0 0 263
8.5-10. . 0 0 0 0 0 . . . . . . . . . . . . . .  0 0 0 0 0 0 0 0 0 0 0 . . 9
8.5-10. . 0 0 0 0 . . . . . . . . . . . . . . .  . . . . . . . . . 0 0 . 9
8.0-10. . . . . 0 0 0 0 . . . . . . . . . . . .  0 0 0 0 . . . . . . . . 7
8.0-10. . . . . . . 0 0 0 0 . . . . . . . . . .  0 0 0 0 . . . . . . . . 4
8.0-10. . . . . 0 0 0 0 . . . . . . . . . . . .  . 0 0 0 0 0 0 0 . . . 31
8.0-10. . . . . 0 0 0 0 0 . . . . . . . . . . .  . . . . . . . . . 0 0 0 31
8.0-10. . . . . . . 0 0 0 0 . . . . . . . . . .  . 0 0 0 0 0 0 . . . 28
8.0-10. . . . . . . 0 0 0 . . . . . . . . . . .  . . . . . . 0 0 0 0 . 28
7.5-10. . . . . . . . 0 0 0 0 0 . . . . . . . .  . . 0 0 0 0 . 52
7.5-10. . . . . . . . . . 0 0 0 0 . . . . . . .  . . . 0 0 0 0 . 49
6.5-10. . . . . . . . . 0 0 0 0 0 . . . . . . .  . 0 0 0 0 0 0 . 26
6.5-10. 0 0 . . . . . . . . . . 0 0 0 0 0 . . .  . . 0 0 0 0 0 0 . 25
6.5-8.5 0 0 . . . . . . . . . . . . . . . . . .  0 0 0 0 0 0 0 0 0 0 0 0 . 264
6.5-8.5 0 0 0 . . . . . . . . . . . . . . . . .  . . . . . . . . . . 0 0 264
6.5-8.5 . 0 0 0 0 . . . . . . . . . . . . . . .  0 0 0 0 0 0 0 0 0 0 0 . 10
6.5-8.5 . 0 0 0 0 . . . . . . . . . . . . . . .  . . . . . . . . . 0 0 . 10
6.5-8.0 . . . . 0 0 0 0 . . . . . . . . . . . .  0 0 0 0 . . . . . . . . 8
6.5-8.0 . . . . . . 0 0 0 0 . . . . . . . . . .  0 0 0 0 . . . . . . . . 5
6.5-10. . . . . . . . 0 0 0 0 0 . . . . . . . .  0 0 0 . . . . . . . . 2
6.5-10. . . . . . . . . . 0 0 0 0 0 0 0 0 . . .  0 0 0 . . . . . . . . 1
6.5-8.0 . . . . 0 0 0 0 . . . . . . . . . . . .  . . 0 0 0 0 0 0 0 . . 32
6.5-8.0 . . . . . . 0 0 0 0 . . . . . . . . . .  . . 0 0 0 0 0 . . . 29
6.5-8.0 . . . . . . 0 0 0 . . . . . . . . . . .  . . 0 0 0 0 . . . 29
6.5-8.0 . . . 0 0 0 0 0 . . . . . . . . . . . .  . . . . . . . 0 0 0 33
6.5-7.5 . . . . . . . 0 0 0 0 . . . . . . . . .  . . . . . 0 0 . . . 53
5.5-6.5 . . . . . . . 0 0 0 0 . . . . . . . . .  . . . . . 0 0 . . . 53
5.5-7.5 . . . . . . . 0 0 0 0 . . . . . . . . .  . . . . 0 0 0 . . 53
5.5-7.5 . . . . . . . . . 0 0 0 0 . . . . . . .  . . . . 0 0 0 0 . 50
5.5-6.5 . . . . . 0 0 0 0 0 . . . . . . . . . .  0 0 0 0 0 . . . . . 18
5.5-6.5 . . . . . 0 0 0 0 . . . . . . . . . . .  . . . 0 0 0 . . . 18
5.5-6.5 . . . . . . 0 0 0 . . . . . . . . . . .  0 0 0 0 . . . . 6
5.5-6.5 . . . . . . . 0 0 0 0 0 0 0 0 . 0 0 0 .  . . . . . . . . 3
5.5-6.5 . . . . . . . 0 0 0 0 0 . . . . . 0 0 .  . . . . . . . . 3
5.5-6.5 . . . . . . . . . 0 0 0 0 . . . 0 0 . .  . . . . . . . . 27
5.5-6.5 . . . . . . . 0 0 0 0 0 0 0 0 . . 0 0 .  . . . . . . . . 27
5.5-6.5 . . . . . . 0 0 0 . . . . . . . 0 0 . .  . . . . . . . . 30
5.5-6.5 0 . . . . . . . . . . . . . . 0 0 . 0 0  0 0 . . . . . . 11
4.5-6.5 0 0 . . . . . . . . . . . . . . . . . .  0 0 0 0 0 0 0 0 0 0 0 . 265
4.5-6.5 0 0 0 . . . . . . . . . . . . . . . . .  . . . . . . . . . . 0 0 265
4.5-6.5 . 0 0 0 0 0 . . . . . . . . . . . . . .  0 0 0 0 0 0 0 . . . . 22
4.5-6.5 . . . . . . . 0 0 0 0 0 . . . . . . . .  0 0 0 0 0 . . . . 37
4.5-6.5 . . . . . . . . . 0 0 . . . . . . . . .  0 0 0 0 0 . . . . 35
4.5-6.5 . . . . . . . . . . . 0 0 0 0 . . . . .  0 0 0 . . . . . . 34
4.5-6.5 . . . . . . 0 0 . . . . . . . . . . . .  0 0 0 0 0 . . . . 42
4.5-6.5 . . . . . . 0 0 0 . . . . . . . . . . .  0 0 0 0 . . . . 39
4.5-6.5 . . . . 0 0 0 0 . . . . . . . . . . . .  . 0 0 0 0 . . . 45
4.5-6.5 . 0 0 0 0 0 . . . . . . . . . . . . . .  . . 0 0 0 . . 63
4.5-6.5 . 0 0 0 0 . . . . . . . . . . . . . . .  . . . . . 0 0 . 63
4.5-6.5 . . 0 0 0 . . . . . . . . . . . . . . .  . . . . . . 0 0 63
4.5-6.5 . . . . . . 0 0 0 . . . . . . . . . . .  . . . . . 0 0 0 0 57
4.5-6.5 . . . . 0 0 0 0 0 . . . . . . . . . . .  . . . . . . 0 0 0 60
```

Table 12.2(cont.)

```
VALUE                      CHROMA                              HUE              NO.
         0 0 0 1 1 1 2 2 3 5 6 7 8 9 1 1 1 1 1 4  9 1 4 6 7 8 9 1 2 3 5 7 8
         0 5 7 0 2 5 0 5             0 1 3 4 5 0  R R R R R R R Y Y Y Y Y Y
                                                  P         R R R R R R
3.5-6.5  . . . . . . . . . . . . . .   0 0 0 0    . . . . . .   0 0 0 . . . . .    34
4.5-5.5  . . . . 0 0 0 0 . . . . . . . . . . . .  . . . . . .   0 0 0 . . . . .    19
4.5-5.5  . . . . . . . . 0 0 0 0 . . . . . . . .  . . . . . . . . 0 0 . . .        54
4.5-5.5  . . . . . . . . 0 0 0 0 0 . . . . . . .  . . . . . . . . . 0 0 0 .        54
4.5-5.5  . . . . . . . . . . 0 0 0 0 . . . . . .  . . . . . . . . . 0 0 0 0        51
4.5-10.  . . . . . . . . . . . . .   0 0 0 . . .  . . . . . . . . . 0 0 0 0        48
3.5-5.5  . . . . 0 0 0 0 0 0 0 . . . . . . . . .  0 0 0 0 . . . . . . . . .        19
3.5-5.5  . . . . . . . . 0 0 0 0 0 . . . . . . .  0 0 0 0 . . . . . . . . .        15
3.5-5.5  . . . . . . . . . . . 0 0 . . . . . . .  0 0 0 0 . . . . . . . . .        12
3.5-5.5  . . . . . . . . . . 0 0 0 0 . 0 0 0 0 .  . . . . . . . . . . . . .        11
3.5-4.5  . . . . . . . . . . . . 0 0 0 0 . . . .  . 0 0 0 . . . . . . . . .        11
3.5-4.5  . . . . . . . . 0 0 0 0 0 . . . . . . .  . 0 0 0 0 . . . . . . . .        38
3.5-4.5  0 0 . . . . . . . . . . 0 0 . . . . . .  . 0 0 0 0 . . . . . . . .        36
2.5-4.5  0 0 . . . . . . . . . . . . . . . . . .  0 0 0 0 0 0 0 0 0 0 0 0 0       266
2.5-4.5  . 0 0 0 0 0 . . . . . . . . . . . . . .  0 0 0 0 0 0 0 . . . . . .        23
2.5-4.5  . 0 0 0 0 0 . . . . . . . . . . . . . .  . . . . . . 0 0 0 0 . . .        64
2.5-4.5  . 0 0 0 0 . . . . . . . . . . . . . . .  . . . . . . . . . 0 0 0 .        64
2.5-4.5  . . . . . . 0 0 0 0 0 0 0 0 0 0 0 0 0 .  . . . . . . . . 0 0 0 0 0        55
2.5-4.5  . . . . 0 0 0 . . . . . . . . . . . . .  . . . . . . . . . . 0 0 .        61
2.5-4.5  . . . . 0 0 0 0 . . . . . . . . . . . .  . . . . . . . . . 0 0 0 0        61
2.5-4.5  . . . . . . 0 0 0 . . . . . . . . . . .  . . . . . . . . . 0 0 0 0        58
2.5-4.5  . . . . 0 0 0 0 . . . . . . . . . . . .  . . . . . . 0 0 0 0 0 . .        46
2.5-4.5  . . . . . 0 0 0 0 . . . . . . . . . . .  . . . . . . 0 0 0 0 . . .        43
2.5-4.5  . . . . 0 0 0 . . . . . . . . . . . . .  . . . . . . . . 0 0 . . .        43
2.5-3.5  . . . . 0 0 0 0 . . . . . . . . . . . .  . . . 0 0 0 . . . . . . .        46
2.5-3.5  . . . . . 0 0 0 0 . . . . . . . . . . .  . . . 0 0 0 . . . . . . .        43
2.5-3.5  . . . . . . . 0 0 0 0 0 0 0 0 0 . . . .  . . . . 0 0 0 . . . . . .        40
2.5-3.5  . . . . 0 0 0 0 . . . . . . . . . . . .  0 0 0 . . . . . . . . . .        20
2.0-2.5  0 0 0 0 0 0 0 0 . . . . . . . . . . . .  0 0 0 . . . . . . . . . .        20
2.0-3.5  . . . . . . 0 0 0 0 0 0 . . . . . . . .  0 0 0 . . . . . . . . . .        16
2.0-3.5  . . . . . . 0 0 0 . . . . . . . . . . .  . 0 0 0 0 . . . . . . . .        16
2.0-3.5  . . . . . . . 0 0 0 . . . . . . . . . .  0 0 0 0 0 . . . . . . . .        13
1.5-2.5  . 0 0 0 0 0 0 0 . . 0 0 . . . . . . . .  . 0 0 0 0 0 0 0 . . . . .        47
1.5-2.5  . . . . . . . 0 0 . . . . . . . . . . .  . 0 0 0 0 0 0 . . . . . .        44
1.5-2.5  . 0 0 0 0 0 0 0 . . . . . . . . . . . .  . . . . . . . . 0 0 0 0 .        62
1.5-2.5  . . . . . . 0 0 0 . . . . . . . . . . .  . . . . . . . . 0 0 0 0 .        59
0.0-3.5  . . . . . . . . . . . . 0 0 0 0 0 . . .  0 0 0 0 0 0 . . . . . . .        11
0.0-2.5  0 0 . . . . . . . . . . . . . . . . . .  0 0 0 0 0 0 0 0 0 0 0 0 0       267
0.0-2.5  . . . . . . . 0 0 0 . . . . . . . . . .  . . . . . . 0 0 0 . . . .        41
0.0-2.5  . . . . . . 0 0 0 0 0 0 0 0 0 0 0 0 . .  . . . . . . 0 0 0 . . . .        41
0.0-2.5  . . . . . . 0 0 0 0 0 0 0 0 0 0 0 0 . .  . . . . . . . . 0 0 0 0 0        56
0.0-2.0  . 0 0 0 . . . . . . . . . . . . . . . .  0 0 0 . . . . . . . . . .        24
0.0-2.0  . . . 0 0 0 0 . . . . . . . . . . . . .  0 0 0 . . . . . . . . . .        21
0.0-2.0  . . . . . . 0 0 0 0 0 . . . . . . . . .  0 0 0 . . . . . . . . . .        17
0.0-2.0  . . . . . . . . 0 0 0 0 0 . . . . . . .  0 0 0 0 0 0 . . . . . . .        14
0.0-1.5  . 0 0 0 . . . . . . . . . . . . . . . .  . 0 0 0 0 0 . . . . . . .        24
0.0-1.5  . . . 0 0 0 0 0 0 . . . . . . . . . . .  . 0 0 0 0 0 0 . . . . . .        44
0.0-1.5  . 0 0 0 . . . . . . . . . . . . . . . .  . . . . . . . . 0 0 0 0 0        65
0.0-1.5  . . . 0 0 0 0 0 0 . . . . . . . . . . .  . . . . . . . . 0 0 0 0 .        59
```

232

Table 12.2 (cont.)

```
VALUE                    CHROMA                              HUE           NO.
         0 0 0 1 1 1 2 2 3 5 6 7 8 9 1 1 1 1 1 4   7 8 1 4 7 9 2 4 8
         . . . . . . . .             0 1 3 4 5 0   Y Y Y Y Y Y Y G G G
         0 5 7 0 2 5 0 5                           R R           Y Y Y

8.5-10.  0 0 0 . . . . . . . . . . . . . . . . .   0 0 0 0 0 0 . . .     263
8.5-10.  0 0 . . . . . . . . . . . . . . . . . .   . . . . . 0 0 0 0     263
8.5-10.  . . 0 0 0 . . . . . . . . . . . . . . .   0 0 . . . . . . .      92
8.5-10.  . . 0 0 0 0 0 . . . . . . . . . . . . .   . 0 0 0 0 0 . . .      92
8.5-10.  . 0 0 0 0 . . . . . . . . . . . . . . .   . . . . . 0 0 0 .      92
8.0-10.  . . . . . . . . 0 0 0 0 0 . . . . . . .   0 0 0 . . . . . .      70
8.0-10.  . . . . . . . . . 0 0 0 0 . . . . . . .   0 0 0 . . . . . .      67
8.0-10.  . . . . 0 0 0 0 . . . . . . . . . . . .   . . 0 0 0 . . . .      89
8.0-10.  . . . . 0 0 0 . . . . . . . . . . . . .   . . . . 0 0 . . .      89
8.0-10.  . . . . . 0 0 0 0 . . . . . . . . . . .   . . 0 0 0 . . . .      86
8.0-10.  . . . . . . . . 0 0 0 0 . . . . . . . .   . . 0 0 0 . . . .      83
8.0-10.  . . . . . . 0 0 . . . . . . . . . . . .   . . . . . 0 0 0 .     104
8.0-10.  . . . . . 0 0 0 0 . . . . . . . . . . .   . . . . . 0 0 0 .     101
8.0-10.  . . . . . . 0 0 0 0 . . . . . . . . . .   . . . . . 0 0 0 .      98
7.5-10.  . . . . . 0 0 0 . . . . . . . . . . . .   0 0 . . . . . . .      73
7.5-10.  . . . . 0 0 0 0 0 . . . . . . . . . . .   . 0 0 . . . . . .      73
7.5-10.  . . . 0 0 0 0 0 . . . . . . . . . . . .   . . . . . 0 0 0 0     121
7.5-10.  . . . . . 0 0 0 0 . . . . . . . . . . .   . . . . . . 0 0 0     119
7.5-10.  . . . . . . 0 0 0 0 . . . . . . . . . .   . . . . . . 0 0 0     116
5.5-10.  . . . . . . . . . . . . . 0 0 0 . . . .   0 0 0 . . . . . .      66
5.5-10.  . . . . . . . . . . . 0 0 0 0 0 . . . .   . . 0 0 0 . . . .      82
5.5-10.  . . . . . . . . . . . 0 0 0 0 0 . . . .   . . . . 0 0 0 . .      97
3.5-10.  . . . . . . . . . . . 0 0 0 0 0 . . . .   . . . . . . 0 0 0     115
6.5-8.5  0 0 0 . . . . . . . . . . . . . . . . .   0 0 0 0 0 0 . . .     264
6.5-8.5  0 0 . . . . . . . . . . . . . . . . . .   . . . . . 0 0 0 0     264
6.5-8.5  . . 0 0 0 . . . . . . . . . . . . . . .   0 0 . . . . . . .      93
6.5-8.5  . . 0 0 0 0 0 . . . . . . . . . . . . .   . 0 0 0 0 0 . . .      93
6.5-8.5  . 0 0 0 0 . . . . . . . . . . . . . . .   . . . . . 0 0 0 .      93
6.5-8.0  . . . . . . . . 0 0 0 0 0 . . . . . . .   0 0 0 . . . . . .      71
6.5-8.0  . . . . . . . . . 0 0 0 0 . . . . . . .   0 0 0 . . . . . .      68
6.5-8.0  . . . . 0 0 0 0 . . . . . . . . . . . .   . . 0 0 0 . . . .      90
6.5-8.0  . . . . 0 0 0 . . . . . . . . . . . . .   . . . . 0 0 . . .      90
6.5-8.0  . . . . . 0 0 0 0 . . . . . . . . . . .   . . 0 0 0 . . . .      87
6.5-8.0  . . . . . . 0 0 0 0 . . . . . . . . . .   . . 0 0 0 . . . .      84
6.5-8.0  . . . . . . 0 0 . . . . . . . . . . . .   . . . . . 0 0 0 .     105
6.5-8.0  . . . . . 0 0 0 0 . . . . . . . . . . .   . . . . . 0 0 0 .     102
6.5-8.0  . . . . . . 0 0 0 0 . . . . . . . . . .   . . . . . 0 0 0 .      99
6.5-7.5  . . . . . 0 0 0 . . . . . . . . . . . .   0 0 . . . . . . .      76
6.5-7.5  . . . . 0 0 0 . . . . . . . . . . . . .   . 0 0 . . . . . .      79
6.5-7.5  . . . 0 0 0 0 0 . . . . . . . . . . . .   . . . . . 0 0 . .     122
5.5-7.5  . . . . . 0 0 0 . . . . . . . . . . . .   . 0 0 . . . . . .      76
5.5-6.5  . . . . . . . . 0 0 0 0 0 . . . . . . .   0 0 0 . . . . . .      72
5.5-6.5  . . . . . . . . . 0 0 0 0 . . . . . . .   0 0 0 . . . . . .      69
5.5-6.5  . . . . 0 0 0 0 0 . . . . . . . . . . .   . . 0 0 . . . . .      79
5.5-6.5  . . . . 0 0 0 0 0 . . . . . . . . . . .   . . 0 0 . . . . .      94
5.5-6.5  . . . . . . 0 0 . . . . . . . . . . . .   . . 0 0 0 . . . .      91
5.5-6.5  . . . . . . . 0 0 0 0 . . . . . . . . .   . . 0 0 0 . . . .      88
5.5-6.5  . . . . . . . . . 0 0 0 0 . . . . . . .   . . 0 0 0 . . . .      85
5.5-6.5  . . . . . . 0 0 . . . . . . . . . . . .   . . . . . 0 0 0 .     106
5.5-6.5  . . . . . . 0 0 0 0 . . . . . . . . . .   . . . . . 0 0 0 .     103
5.5-6.5  . . . . . . . 0 0 0 0 . . . . . . . . .   . . . . . 0 0 0 .     100
```

Table 12.2(cont.)

```
VALUE                      CHROMA                              HUE          NO.
          0 0 0 1 1 1 2 2 3 5 6 7 8 9 1 1 1 1 1 4.   7 8 1 4 7 9 2 4 8
          . . . . . . . .               0 1 3 4 5 0  Y Y Y Y Y Y G G G
          0 5 7 0 2 5 0 5                             R R       Y Y Y

4.5-7.5   . . . . 0 0 0 0 0 . . . . . . . . . . . .   . . . . . . 0 0 0    122
4.5-7.5   . . . . . . 0 0 0 0 . . . . . . . . . . .   . . . . . . 0 0 0    120
4.5-7.5   . . . . . . . . 0 0 0 0 0 . . . . . . . .   . . . . . . 0 0 0    117
4.5-6.5   0 0 0 . . . . . . . . . . . . . . . . . .   0 0 0 0 0 0 . . .    265
4.5-6.5   0 0 . . . . . . . . . . . . . . . . . . .   . . . . . 0 0 0 0    265
4.5-6.5   . . 0 0 0 . . . . . . . . . . . . . . . .   0 0 0 0 . . . . .     63
4.5-6.5   . . 0 0 0 0 0 . . . . . . . . . . . . . .   . . . 0 0 0 . . .    112
4.5-6.5   . 0 0 0 0 . . . . . . . . . . . . . . . .   . . . . . 0 0 0 0    112
4.5-6.5   . . . . . 0 0 0 . . . . . . . . . . . . .   . . . 0 0 0 . . .    109
4.5-6.5   . . . . 0 0 0 0 0 . . . . . . . . . . . .   . . . . 0 0 . . .    109
4.5-5.5   . . . . . . . . . 0 0 0 0 0 0 0 0 0 0 . .   0 0 . . . . . . .     74
4.5-5.5   . . . . 0 0 0 0 0 . . . . . . . . . . . .   . 0 0 . . . . . .     80
4.5-5.5   . . . . . . . 0 0 . . . . . . . . . . . .   . 0 0 . . . . . .     77
4.5-5.5   . . . . 0 0 0 0 0 0 0 0 0 0 0 0 0 0 0 0 .   . . 0 0 . . . . .     94
4.5-5.5   . . . . . . . . 0 0 0 0 0 0 0 0 0 0 . . .   . 0 0 . 0 0 0 . .    106
3.5-5.5   . . . . . . . . 0 0 0 0 0 0 0 0 0 0 . . .   . 0 0 . . . . . .     74
3.5-4.5   . . . 0 0 0 0 . . . . . . . . . . . . . .   . 0 0 . . . . . .     80
3.5-4.5   . . . . . . 0 0 0 . . . . . . . . . . . .   . 0 0 . . . . . .     77
3.5-4.5   . . . . . . . . . 0 0 0 0 0 . . . . . . .   . . . . . . 0 0 0    118
2.5-4.5   0 0 . . . . . . . . . . . . . . . . . . .   0 0 0 0 0 0 0 0 0    266
2.5-4.5   . 0 0 0 0 . . . . . . . . . . . . . . . .   0 0 0 0 . . . . .     64
2.5-4.5   . . . 0 0 0 0 0 0 0 0 0 0 0 0 0 0 0 0 0 .   . . 0 0 . . . . .     95
2.5-4.5   . 0 0 0 0 0 . . . . . . . . . . . . . . .   . . . 0 0 0 . . .    113
2.5-4.5   . 0 0 0 0 . . . . . . . . . . . . . . . .   . . . . . 0 0 0 .    113
2.5-4.5   . . . . 0 0 0 0 . . . . . . . . . . . . .   . . . 0 0 0 . . .    110
2.5-4.5   . . . . 0 0 0 0 0 . . . . . . . . . . . .   . . . . 0 0 . . .    110
2.5-4.5   . . . . . . 0 0 0 0 0 0 0 0 0 0 0 0 . . .   . . . 0 0 0 0 . .    107
2.5-4.5   . . . . 0 0 0 0 0 . . . . . . . . . . . .   . . . . . . 0 0 0    127
2.5-4.5   . . . . . . 0 0 0 0 . . . . . . . . . . .   . . . . . . 0 0 0    125
2.5-3.5   . . . . . . . . 0 0 0 0 0 0 0 0 0 0 . . .   . . . . . . 0 0 0    123
2.5-3.5   . . . . 0 0 0 0 . . . . . . . . . . . . .   . 0 0 . . . . . .     81
1.5-3.5   . . . . . . . 0 0 0 . . . . . . . . . . .   . 0 0 . . . . . .     78
1.5-2.5   . 0 0 0 0 0 0 0 . . . . . . . . . . . . .   . 0 0 . . . . . .     81
1.5-2.5   . 0 0 0 0 0 0 0 0 0 0 0 0 0 0 0 0 0 0 0 .   . . 0 0 . . . . .     96
1.5-2.5   . 0 0 0 0 0 0 0 . . . . . . . . . . . . .   . . . 0 0 0 0 . .    111
1.5-2.5   . . . . . . . 0 0 0 0 0 0 0 0 0 0 0 0 . .   . . . 0 0 0 0 . .    108
1.5-2.5   . 0 0 0 0 0 0 0 . . . . . . . . . . . . .   . . . . . . 0 0 0    128
1.5-2.5   . . . . . . . 0 0 0 0 . . . . . . . . . .   . . . . . . 0 0 0    126
0.0-3.5   . . . . . . . . 0 0 0 0 0 0 0 0 0 0 0 . .   . 0 0 . . . . . .     75
0.0-2.5   . . . . . . . . . 0 0 0 0 0 0 0 0 0 . . .   . . . . . . 0 0 0    124
0.0-1.5   . . . 0 0 0 0 0 0 0 . . . . . . . . . . .   . 0 0 . . . . . .     78
0.0-1.5   . 0 0 0 . . . . . . . . . . . . . . . . .   0 0 0 0 . . . . .     65
0.0-1.5   . . . 0 0 0 0 0 0 0 0 0 0 0 0 0 0 0 0 0 .   . . 0 0 . . . . .     96
0.0-1.5   . . . 0 0 0 0 0 0 0 0 0 0 0 0 0 0 0 0 0 .   . . . 0 0 0 0 . .    108
0.0-1.5   . 0 0 0 . . . . . . . . . . . . . . . . .   . . . 0 0 0 0 0 .    114
0.0-1.5   . . . 0 0 0 0 0 0 0 0 0 . . . . . . . . .   . . . . . . 0 0 0    126
0.0-2.5   0 0 . . . . . . . . . . . . . . . . . . .   0 0 0 0 0 0 0 0 0    267
```

Table 12.2(cont.)

VALUE	0.0	0.5	0.7	1.0	1.2	1.5	2.0	2.5	3	5	6	7	8	9	10	11	13	14	15	40	4GY	8GY	3G	9G	10BG	9B	5PB	6PB	7PB	9PB	NO.
8.5-10.	O	O																			O	O	O	O	O	O	O	O	O	O	263
8.5-10.		O	O	O	O																O	O	O	O	O						153
8.5-10.								O	O	O	O	O										O	O								134
8.5-10.		O	O	O	O	O																				O	O	O	O	O	189
7.5-10.				O	O	O	O															O	O	O	O						148
7.5-10.							O	O	O	O	O												O	O							143
7.5-10.							O	O	O	O	O													O	O						162
7.5-10.						O	O	O	O																O	O					184
7.5-10.						O	O	O	O	O															O	O					184
7.5-10.						O	O	O	O																	O	O	O	O		184
7.5-10.								O	O	O	O														O	O					171
7.5-10.									O	O	O	O	O													O	O				180
7.5-10.										O	O	O														O	O	O			180
7.5-10.							O	O																			O	O	O	O	202
7.5-10.									O	O	O																O	O	O		198
7.5-10.									O	O	O	O	O																O	O	198
6.5-10.									O	O	O	O	O										O	O							130
5.5-10.										O	O	O	O	O										O	O						140
5.5-10.										O	O	O	O	O										O	O						159
5.5-10.										O	O	O	O	O											O	O					168
5.5-10.											O	O	O	O													O	O	O	O	177
5.5-10.												O	O	O	O														O	O	195
4.5-10.												O	O	O	O	O							O	O							129
3.0-10.													O	O	O	O	O										O	O	O	O	176
3.0-10.													O	O	O	O	O												O	O	194
0.0-10.													O	O	O	O	O							O	O						139
0.0-10.													O	O	O	O	O							O	O						158
0.0-10.													O	O	O	O	O								O	O					167
6.5-8.5		O	O	O	O																O	O	O	O							154
6.5-8.5							O	O	O	O	O												O	O							135
6.5-8.5			O	O	O	O																				O	O	O	O	O	190
6.5-8.5	O	O																			O	O	O	O	O	O	O	O	O	O	264
5.5-7.5					O	O	O	O															O	O	O						149
5.5-7.5							O	O	O	O	O													O	O						144
5.5-7.5							O	O	O	O	O														O	O					163
5.5-7.5						O	O	O	O																	O	O				185
5.5-7.5						O	O	O	O																		O	O			185
5.5-7.5						O	O	O	O																		O	O	O	O	185
5.5-7.5								O	O	O	O															O	O				172
5.5-7.5										O	O	O	O														O	O			181
5.5-7.5											O	O	O														O	O	O		181
4.5-7.5								O	O																		O	O	O	O	203
4.5-7.5									O	O	O																		O	O	199
4.5-7.5									O	O	O	O	O									O	O								136
4.5-6.5											O	O	O	O	O							O	O								131
4.5-6.5		O	O	O	O																O	O	O	O							155
4.5-6.5		O	O	O	O																					O	O	O	O	O	191
4.5-6.5	O	O																			O	O	O	O	O	O	O	O	O	O	265
4.5-5.5												O	O	O														O	O	O	182
4.5-5.5													O	O	O														O	O	195
4.5-5.5														O	O														O	O	196
3.5-5.5						O	O	O	O	O																	O	O			145
3.5-5.5								O	O	O	O	O															O	O			141
3.5-5.5				O	O	O	O															O	O	O	O						150

Table 12.2(cont.)

VALUE	\.\.\.\. CHROMA \.\.\.\.																			HUE										NO.
	0.0	0.5	0.7	1.0	1.2	1.5	2.0	2.5	3	5	6	7	8	9	11.0	11.3	11.4	11.5	14.0	4GY	8GY	3G	9G	10BG	9B	5PB	6PB	7PB	9PB	
3.5–5.5	.	.	.	.	.	.	0	0	0	0	0	.	.	.	.	.	.	.	.	.	.	.	0	0	.	.	.	.	.	164
3.5–5.5	.	.	.	.	.	.	.	.	0	0	0	0	0	.	.	.	.	.	.	.	.	.	0	0	.	.	.	.	.	160
3.5–5.5	.	.	.	.	.	.	0	0	0	0	.	.	.	.	.	.	.	.	.	.	.	.	.	0	0	.	.	.	.	173
3.5–5.5	.	.	.	.	.	.	.	0	0	0	0	.	.	.	.	.	.	.	.	.	.	.	.	0	0	.	.	.	.	169
3.0–5.5	.	.	.	.	.	0	0	0	0	.	.	.	.	.	.	.	.	.	.	.	.	.	.	0	0	.	.	.	.	186
3.0–5.5	.	.	.	.	.	0	0	0	0	0	.	.	.	.	.	.	.	.	.	.	.	.	.	.	0	0	.	.	.	186
3.0–5.5	.	.	.	.	.	0	0	0	0	.	.	.	.	.	.	.	.	.	.	.	.	.	.	.	0	0	0	0	.	186
3.0–5.5	.	.	.	.	.	.	.	.	0	0	0	0	0	.	.	.	.	.	.	.	.	.	.	.	0	0	.	.	.	182
3.0–5.5	.	.	.	.	.	.	.	.	.	0	0	0	0	.	.	.	.	.	.	.	.	.	.	.	0	0	0	0	.	178
3.0–4.5	.	.	.	.	.	.	.	0	0	0	0	0	.	.	.	.	.	.	.	.	.	.	.	.	0	0	0	.	.	182
3.0–4.5	.	.	.	.	.	.	.	0	0	.	.	.	.	.	.	.	.	.	.	.	.	.	.	.	0	0	.	.	.	204
3.0–4.5	.	.	.	.	.	.	.	.	.	0	0	0	0	.	.	.	.	.	.	.	.	.	.	.	.	0	0	.	.	196
3.0–4.5	.	.	.	.	.	.	.	0	0	0	0	0	.	.	.	.	.	.	.	.	.	.	.	.	.	0	0	.	.	200
2.5–4.5	.	.	.	.	.	.	0	0	0	0	0	.	.	.	.	.	.	.	.	.	0	0	.	.	.	.	.	.	.	137
2.5–4.5	.	.	.	.	.	.	.	.	0	0	0	0	0	0	0	0	0	.	.	.	0	0	.	.	.	.	.	.	.	132
2.5–4.5	.	.	.	.	.	.	.	0	0	.	.	.	.	.	.	.	.	.	.	.	.	.	.	.	0	0	0	.	.	204
2.5–4.5	.	0	0	0	0	.	.	.	.	.	.	.	.	.	.	.	.	.	.	0	0	0	0	0	.	.	.	.	.	156
2.5–4.5	.	0	0	0	0	0	.	.	.	.	.	.	.	.	.	.	.	.	.	.	.	.	.	0	0	0	0	0	.	192
2.5–4.5	0	0	.	.	.	.	.	.	.	.	.	.	.	.	.	.	.	.	.	0	0	0	0	0	0	0	0	0	0	266
2.5–3.5	.	.	.	.	0	0	0	0	.	.	.	.	.	.	.	.	.	.	.	.	0	0	0	.	.	.	.	.	.	151
2.5–3.5	.	.	.	.	.	.	0	0	0	0	0	.	.	.	.	.	.	.	.	.	.	0	0	.	.	.	.	.	.	146
2.5–3.5	.	.	.	.	.	.	0	0	0	0	.	.	.	.	.	.	.	.	.	.	.	.	0	0	.	.	.	.	.	165
2.5–3.5	.	.	.	.	.	.	0	0	0	0	.	.	.	.	.	.	.	.	.	.	.	.	.	0	0	.	.	.	.	174
2.5–3.0	.	.	.	.	.	0	0	0	0	.	.	.	.	.	.	.	.	.	.	.	.	.	.	0	0	0	0	0	.	187
2.5–3.0	.	.	.	.	.	.	0	0	0	0	.	.	.	.	.	.	.	.	.	.	.	.	.	.	0	0	0	.	.	183
2.0–3.0	.	.	.	.	.	.	0	0	0	.	.	.	.	.	.	.	.	.	.	.	.	.	.	.	.	0	0	.	.	200
2.0–2.5	.	0	0	0	0	0	0	.	.	.	.	.	.	.	.	.	.	.	.	.	0	0	0	.	.	.	.	.	.	151
2.0–2.5	.	.	.	.	.	0	0	0	0	0	.	.	.	.	.	.	.	.	.	.	.	0	0	.	.	.	.	.	.	146
2.0–2.5	.	.	.	.	.	0	0	0	0	0	.	.	.	.	.	.	.	.	.	.	.	0	0	.	.	.	.	.	.	165
2.0–2.5	.	0	0	0	0	0	0	.	.	.	.	.	.	.	.	.	.	.	.	.	.	.	.	0	0	0	0	0	.	187
2.0–2.5	.	.	.	.	.	0	0	0	0	0	.	.	.	.	.	.	.	.	.	.	.	.	0	0	.	.	.	.	.	174
2.0–2.5	.	.	.	.	.	0	0	0	0	.	.	.	.	.	.	.	.	.	.	.	.	.	.	0	0	0	.	.	.	204
0.0–3.5	.	.	.	.	.	.	.	.	0	0	0	0	0	.	.	.	.	.	.	.	.	0	0	.	.	.	.	.	.	142
0.0–3.5	.	.	.	.	.	.	.	.	0	0	0	0	0	.	.	.	.	.	.	.	.	0	0	.	.	.	.	.	.	161
0.0–3.5	.	.	.	.	.	.	.	.	0	0	0	0	0	.	.	.	.	.	.	.	.	0	0	.	.	.	.	.	.	170
0.0–3.0	.	.	.	.	.	.	.	0	0	0	.	.	.	.	.	.	.	.	.	.	.	.	.	.	0	0	.	.	.	183
0.0–3.0	.	.	.	.	.	.	.	.	0	0	0	0	.	.	.	.	.	.	.	.	.	.	.	.	0	0	0	0	.	179
0.0–3.0	.	.	.	.	.	.	.	.	.	.	0	0	0	0	.	.	.	.	.	.	.	.	.	.	0	0	0	0	.	176
0.0–3.0	.	.	.	.	.	.	.	.	.	.	0	0	0	0	.	.	.	.	.	.	.	.	.	.	.	.	0	0	.	194
0.0–3.0	.	.	.	.	.	.	.	.	0	0	0	0	.	.	.	.	.	.	.	.	.	.	.	.	.	.	0	0	.	197
0.0–2.5	.	.	.	.	.	0	0	0	0	0	0	.	.	.	.	.	.	.	.	.	0	0	.	.	.	.	.	.	.	138
0.0–2.5	.	.	.	.	.	.	.	0	0	0	0	0	0	0	0	.	.	.	.	.	0	0	.	.	.	.	.	.	.	133
0.0–2.5	.	.	.	.	.	0	0	0	0	0	0	.	.	.	.	.	.	.	.	.	.	.	.	.	0	0	0	.	.	183
0.0–2.5	0	0	.	.	.	.	.	.	.	.	.	.	.	.	.	.	.	.	.	0	0	0	0	0	0	0	0	0	0	267
0.0–2.0	.	.	.	.	.	.	0	0	0	0	0	.	.	.	.	.	.	.	.	.	.	.	0	0	.	.	.	.	.	147
0.0–2.0	.	.	.	.	.	.	0	0	0	0	0	.	.	.	.	.	.	.	.	.	.	.	0	0	.	.	.	.	.	166
0.0–2.0	.	.	.	.	.	.	0	0	0	0	0	.	.	.	.	.	.	.	.	.	.	.	0	0	.	.	.	.	.	175
0.0–2.0	.	.	.	.	.	.	0	0	0	0	.	.	.	.	.	.	.	.	.	.	.	.	.	.	.	.	0	0	.	201
0.0–2.0	.	.	.	.	.	.	0	0	0	0	0	.	.	.	.	.	.	.	.	.	.	.	.	.	.	.	0	0	.	201
0.0–2.0	.	0	0	0	.	.	.	.	.	.	.	.	.	.	.	.	.	.	.	.	.	.	.	0	0	0	0	0	.	193
0.0–2.0	.	.	0	0	0	0	.	.	.	.	.	.	.	.	.	.	.	.	.	.	.	.	0	0	0	0	0	0	.	188
0.0–2.0	.	.	0	0	0	0	0	.	.	.	.	.	.	.	.	.	.	.	.	.	0	0	0	0	.	.	.	.	.	152
0.0–2.0	.	0	0	0	.	.	.	.	.	.	.	.	.	.	.	.	.	.	.	.	0	0	0	0	.	.	.	.	.	157
0.0–1.5	.	0	0	0	.	.	.	.	.	.	.	.	.	.	.	.	.	.	.	0	0	.	.	.	.	.	.	.	.	157

Table 12.2(cont.)

```
VALUE                      CHROMA                        HUE       NO.

         0 0 0 1 1 2 2 3 5 6 7 8 9 1 1 1 1 4    9 3 9 3 9 1
         . . . . . . . .             0 1 3 4 5 0    P P P R R R
         0 5 7 0 2 5 0 5                           B   P P

8.5-10.  O O . . . . . . . . . . . . . . . .    O O O O O O   263
8.5-10.  . O O O O O . . . . . . . . . . . .    O O O O O .   231
7.5-10.  . . . . . O O O O . . . . . . . . .    O O . . . .   226
7.5-10.  . . . . . O O O O O . . . . . . . .    . O O . . .   226
7.5-10.  . . . . . . . O O . . . . . . . . .    O O . . . .   213
7.5-10.  . . . . . . O O O O O . . . . . . .    O O . . . .   209
7.5-10.  . . . . . . O O O O O . . . . . . .    . O O . . .   221
7.5-10.  . . . . . O O O O O . . . . . . . .    . . O O O .   252
7.5-10.  . . . . . . O O O O O . . . . . . .    . . O O O .   249
7.5-10.  . . . . . . . . . O O O O O O O . .    . . O O O .   246
5.5-10.  . . . . . . . . . O O O O . . . . .    . O O . . .   217
4.5-10.  . . . . . . . . . O O O O . . . . .    O O . . . .   206
0.0-10.  . . . . . . . . . . . O O O O . . .    O O . . . .   205
0.0-10.  . . . . . . . . . . . O O O O . . .    . O O . . .   216
6.5-8.5  . O O O O O . . . . . . . . . . . .    O O O O O .   232
6.5-8.5  O O . . . . . . . . . . . . . . . .    O O O O O O   264
6.5-7.5  . . . . . O O O O O . . . . . . . .    . . O O O .   253
6.5-7.5  . . . . . . . O O O O O . . . . . .    . . O O O .   250
6.5-7.5  . . . . . . . . . O O O O O O O . .    . . O O O .   247
5.5-7.5  . . . . . O O O O . . . . . . . . .    O O . . . .   227
5.5-7.5  . . . . . O O O O O . . . . . . . .    . O O . . .   227
5.5-7.5  . . . . . . . O O O O O . . . . . .    . O O . . .   222
4.5-7.5  . . . . . . . O O . . . . . . . . .    O O . . . .   214
4.5-7.5  . . . . . . . O O O O O . . . . . .    O O . . . .   210
4.5-6.5  . O O O O O . . . . . . . . . . . .    O O O O O O   233
4.5-6.5  O O . . . . . . . . . . . . . . . .    O O O O O O   265
5.5-6.5  . . . . . O O O O . . . . . . . . .    . . O O O O   227
5.5-6.5  . . . . . . . O O . . . . . . . . .    . . O O . .   244
5.5-6.5  . . . . . . . O O O O O . . . . . .    . . O O . .   240
5.5-6.5  . . . . . . . . . O O O O O O . . .    . . O O O .   248
5.5-6.5  . . . . . . . . . . . . . O O . . .    . . O O . .   236
5.5-6.5  . . . . . . . O O . . . . . . . . .    . . O O O O   261
5.5-6.5  . . . . . . . O O O O O . . . . . .    . . O O O .   251
5.5-6.5  . . . . . . . . . . . . . O O . . .    . . O O O O   254
3.5-5.5  . . . . . O O O O . . . . . . . . .    O O . . . .   228
3.5-5.5  . . . . . O O O O O . . . . . . . .    . O O . . .   228
3.5-5.5  . . . . . O O O O . . . . . . . . .    . . O O O O   228
3.5-5.5  . . . . . . . O O O O O . . . . . .    . . O O . .   223
3.5-5.5  . . . . . . . . . O O O O . . . . .    . . O O . .   218
3.5-5.5  . . . . . . . O O . . . . . . . . .    . . O O . .   245
3.5-5.5  . . . . . . . O O O O O . . . . . .    . . O O . .   241
3.5-5.5  . . . . . . . . . O O O O . . . . .    . . O O . .   237
3.5-5.5  . . . . . . O O O O . . . . . . . .    . . . O O O   262
3.5-5.5  . . . . . . . O O O O O . . . . . .    . . . O O O   258
3.5-5.5  . . . . . . . . . . . O O . . . . .    . . . O O O   255
2.5-4.5  . . . . . . . O O . . . . . . . . .    O O . . . .   215
2.5-4.5  . . . . . . . O O O O O . . . . . .    O O . . . .   211
2.5-4.5  . . . . . . . . . O O O O . . . . .    O O . . . .   207
2.5-4.5  . O O O O O . . . . . . . . . . . .    O O O O O O   234
2.5-4.5  O O . . . . . . . . . . . . . . . .    O O O O O O   266
```

Table 12.2(cont.)

```
VALUE                   CHROMA                              HUE        NO.
          0 0 0 1 1 1 2 2 3 5 6 7 8 9 1 1 1 1 4    9 3 9 3 9 1
          . . . . . . . .             0 1 3 4 5    P P P R R R
          0 5 7 0 2 5 0 5             0 1 3 4 5 0   B     P P

2.5-3.5   . . . . . 0 0 0 0 . . . . . . . . . . .  0 0 0 0 0 0   229
2.5-3.5   . . . . . . 0 0 0 0 . . . . . . . . . .  . 0 0 . . .   224
2.5-3.5   . . . . . . . 0 0 0 0 . . . . . . . . .  . . 0 0 . .   242
2.5-3.5   . . . . . . . 0 0 0 0 0 0 . . . . . . .  . . . 0 0 0   259
2.0-5.5   . . . . . . . . . . . . . . 0 0 0 0 . .  . . . 0 0 0   254
2.0-3.5   . . . . . . . . . 0 0 0 0 0 0 . . . . .  . 0 0 . . .   219
2.0-3.5   . . . . . . . . . 0 0 0 0 0 0 . . . . .  . . 0 0 . .   238
2.0-3.5   . . . . . . . . . . . 0 0 0 0 . . . . .  . . . 0 0 0   256
2.0-2.5   . 0 0 0 0 0 . . . . . . . . . . . . . .  0 0 0 0 0 0   229
2.0-2.5   . . . . . 0 0 0 0 0 0 . . . . . . . . .  . 0 0 . . .   224
2.0-2.5   . . . . . 0 0 0 0 0 0 . . . . . . . . .  . . 0 0 . .   242
2.0-2.5   . . . . . 0 0 0 0 0 0 0 0 . . . . . . .  . . . 0 0 0   259
0.0-5.5   . . . . . . . . . . . . . . 0 0 0 0 . .  . . 0 0 . .   236
0.0-2.5   . . . . . 0 0 0 0 0 0 . . . . . . . . .  0 0 . . . .   212
0.0-2.5   . . . . . . . . . 0 0 0 0 0 0 . . . . .  0 0 . . . .   208
0.0-2.5   0 0 . . . . . . . . . . . . . . . . . .  0 0 0 0 0 0   267
0.0-2.0   . 0 0 0 . . . . . . . . . . . . . . . .  0 0 0 0 0 0   235
0.0-2.0   . . . 0 0 0 0 . . . . . . . . . . . . .  0 0 0 0 0 0   230
0.0-2.0   . . . . . 0 0 0 0 0 0 . . . . . . . . .  . 0 0 . . .   225
0.0-2.0   . . . . . . . . . 0 0 0 0 0 0 . . . . .  . 0 0 . . .   220
0.0-2.0   . . . . . 0 0 0 0 0 0 . . . . . . . . .  . . 0 0 . .   243
0.0-2.0   . . . . . . . . . 0 0 0 0 0 . . . . . .  . . 0 0 . .   239
0.0-2.0   . . . . . . . . . . . . . . 0 0 0 0 0 .  . . . 0 0 0   254
0.0-2.0   . . . . . 0 0 0 0 0 0 . . . . . . . . .  . . . 0 0 0   260
0.0-2.0   . . . . . . . . . 0 0 0 0 0 . . . . . .  . . . 0 0 0   257
```

12.4 Notes

Detailed technical discussions and mathematical equations have not been included in the text of this book in order to make it understandable by those who do not have a technical or scientific background. This Notes section has been added for those readers wishing some further technical comment and convenient access to some useful equations and supplementary information.

2.1 Since what we see at low levels of illumination, especially in moonlight, often appears bluish, some portion of the visual system associated with perception of blue may be stimulated.

3.1 The definition given is based on the CIE (1970) definition: "Saturation – attribute of a visual sensation which permits a judgment to be made of the proportion of pure chromatic color in the total sensation. This attribute is the psychosensorial correlate, or nearly so, of the colorimetric quantity purity" [3.1,3].

3.2 For purposes of unambiguity and to avoid confusion with the widely accepted term "Munsell Chroma", Hunt [3.3] has proposed the term "colorfulness" to replace the term "chroma" proposed by the CIE [3.1] for the attribute.

4.1 Physicists often use the term "light" for electromagnetic radiation in the wavelength range that includes not only visible radiation but also infrared and ultraviolet radiation and x-rays.

4.2 The relative spectral power distribution curves for CIE ILL B and CIE ILL C (Sect. 4.6) accepted by the CIE in 1931 do not pass through the point corresponding to 100 at wavelength 560 nm, but their deviations from the present convention are minor.

5.1 A spectral reflectance curve (or a spectral transmittance curve) is an accurate indicator of the wavelength composition of light scattered by an opaque material (or transmitted by a transparent material) if it can be assumed that an equal-energy light source provided the illumination [Ref. 2.5, p. 52].

5.2 This behavior is defined mathematically by Bouguer's law (sometimes called Lambert's law) of absorption in homogeneous materials [Ref. 4.4, p. 39].

6.1 1) Calculation of *CIE chromaticity* (x, y) when the tristimulus values X, Y, and Z are given

$$x = \frac{X}{X + Y + Z} \qquad y = \frac{Y}{X + Y + Z} \ .$$

2) Calculation of the *CIE tristimulus values* when x, y, and Y are given

$$X = \frac{x}{y}Y \qquad Z = \frac{(1 - x - y)}{y}Y \ .$$

6.2 1) Calculation of *CIE 1960 chromaticity* (u, v) and *CIE 1976 chromaticity* (u', v') when the tristimulus values X, Y, and Z, or x, y, and Y are given

$$u = u' = \frac{4X}{X + 15Y + 3Z} = \frac{4x}{12y - 2x + 3}$$

$$v = \frac{6Y}{X + 15Y + 3Z} = \frac{6y}{12y - 2x + 3}$$

$$v' = \frac{9Y}{X + 15Y + 3Z} = \frac{9y}{12y - 2x + 3} \ .$$

2) Calculation of *CIE chromaticity* (x, y) when CIE 1960 chromaticity (u, v) or CIE 1976 chromaticity (u', v') is given

$$x = \frac{3u}{2u - 8v + 4} = \frac{9u'}{6u' - 16v' + 12}$$

$$y = \frac{2v}{2u - 8v + 4} = \frac{4v'}{6u' - 16v' + 12} \ .$$

7.1 The representation of black at *any* point on the chromaticity diagram at $Y = 0$ can be demonstrated by the trend indicated in the series of Munsell-CIE conversion charts from $Y = 9$ to $Y = 1$ (Figs. 12.9 to 12.1) and from $Y = 0.8$ to $Y = 0.2$ in [Ref. 7.6, Figs. 1–4]. There we can see the accelerating growth of the achromatic zone (defined, say, by Chroma=1, see Sect. 8.4); the neutral grays get blacker and the Chroma=1 zone expands as $Y = 0$ is approached.

The maximum number of equally spaced colors at any lightness level L^* in CIELAB color space (Sect. 8.2), in which the uniformity of spacing of perceived colors (Sect. 7.10) may be the best presently available, can be assumed to be proportional to the area within the MacAdam limit at that level. The areas measured in [Ref. 8.2, Figs. 3,4] for $L^* = 70, 60, 50$, and 30

are approximately 2.6, 2.9, 3.1, and $2.0\,\mathrm{cm}^2$, respectively. A plot of these data suggests that the maximum number of equally spaced colors are found at a level in the neighborhood of $L^* = 50$.

The model of CIELUV color space (Sect. 8.2) in Fig. 8.8 also shows the maximum cross-sectional area at a level near $L^* = 50$ (or about $Y = 0.2$, see Table 8.2). The color black is represented by the point $L^* = 0$, at the bottom tip.

7.2 A beam of ordinary light (e.g., daylight, lamplight) can be considered to consist of two types of circularly polarized light (i.e., right- and left-handed). But only one type (composing 50% of the incident beam) can interact with a given cholesteric liquid crystal to produce selective reflection. Thus the maximum value of the reflectance is 0.50. In analogy with MacAdam's theoretical approach employed in the case of ordinary nonfluorescent colorants for which the maximum reflectance is 1.00, Makow found that the chromaticity limits for the colors of cholesteric liquid crystals may be obtained from the MacAdam limits (Fig. 7.6) by reducing the values of Y indicated by the latter limits by 50% [7.16].

7.3 Calculation of the *CIE (1976) lightness function L^**, also called the *metric lightness*.

$$L^* = 116\left(\frac{Y}{Y_0}\right)^{1/3} - 16 \ , \quad (Y/Y_0 > 0.01)$$

where, in the case of object colors, Y/Y_0 is the luminance factor (applicable for a given illuminant).

Here Y/Y_0 is the luminance factor for a surface in a given illumination, where the tristimulus value Y [or, equivalently, luminance (Sect. 6.3)] refers to the color of the surface with reference to the illuminant, and the tristimulus value Y_0 (or luminance) refers to a reference white illuminated by the same illuminant. Because the reference white is usually taken to be a perfectly reflecting diffuser, for which $Y_0 = 100$ [Note 7.4], the symbol Y itself is commonly used to represent the luminance factor. This convention is followed in this book (e.g., Table 8.2). (For the case of self-luminous objects, see [7.35].)

The lightness function L^* for a color sample correlates approximately uniformly with perceived lightness [and Munsell Value (CIE ILL C)] under conditions where the surround is intermediate between white-appearing and dark [12.4]. See Table 8.2 for conversions between V, Y (luminance factor), and L^*.

7.4 For the *calculation of the difference between two colors*, two formulas recommended by the CIE are presently available. The following information is required for their use:

a) For illuminated objects: An identification of the illuminant, for example CIE ILL D_{65}.

b) The tristimulus values X_0, Y_0, and Z_0 of the reference white (usually considered a perfect reflecting diffuser, $Y_0 = 100$) illuminated by the given illuminant; see Table 12.3.

c) A set of tristimulus values X, Y, and Z for each color.

d) The CIE (1976) lightness function L^* for each color [Note 7.3].

Table 12.3. The tristimulus values of the reference white illuminated by four different illuminants. [Ref. 1.18, pp. 472, 474, 476, 478]

Illuminant	Tristimulus values (Sect. 6.4)					
	X_0	Y_0	Z_0	$(X_0)_{10}$	$(Y_0)_{10}$	$(Z_0)_{10}$
CIE ILL C	98.04	100.00	118.10	97.30	100.00	116.14
CIE ILL D_{65}	95.02	100.00	108.81	94.83	100.00	107.38
CIE ILL A	109.83	100.00	35.55	111.16	100.00	35.20
E	100.00	100.00	100.00	100.00	100.00	100.00

1) Calculation of the color difference ΔE, by use of the *CIELAB (1976) color-difference formula*

$$\Delta E(\text{LAB}) = \sqrt{(\Delta L^*)^2 + (\Delta a^*)^2 + (\Delta b^*)^2} \ ,$$

where ΔL^*, Δa^*, and Δb^* represent the differences of L^*, a^*, and b^*, respectively, for the two colors, and

$$a^* = 500 \left[\left(\frac{X}{X_0} \right)^{1/3} - \left(\frac{Y}{Y_0} \right)^{1/3} \right] , \quad b^* = 200 \left[\left(\frac{Y}{Y_0} \right)^{1/3} - \left(\frac{Z}{Z_0} \right)^{1/3} \right] .$$

This formula is valid only when X/X_0, Y/Y_0, and Z/Z_0 all exceed 0.01. (When $Y/Y_0 = 0.01$, $L^* = 8.9914$ and the Munsell Value $V = 0.9$, approximately.)

2) Calculation of the color difference ΔE, by use of the *CIELUV (1976) color-difference formula*

$$\Delta E(\text{LUV}) = \sqrt{(\Delta L^*)^2 + (\Delta u^*)^2 + (\Delta v^*)^2} \ ,$$

where ΔL^*, Δu^*, and Δv^* represent the differences of L^*, u^*, and v^*, respectively, for the two colors, and $u^* = 13L^*(u' - u_0')$; $v^* = 13L^*(v' - v_0')$, with

$$u' = \frac{4X}{X + 15Y + 3Z} \; ; \quad v' = \frac{9Y}{X + 15Y + 3Z} \; .$$

See Table 12.4 for values of u'_0 and v'_0 for a reference white illuminated by a standard illuminant.

For the application of the CIELAB and CIELUV formulas in the design of self-luminous displays see [7.35].

Table 12.4. Values of the CIE 1976 (u'_0, v'_0) und CIE 1976 $((u'_0)_{10}, (v'_0)_{10})$ chromaticity coordinates for four different illuminants

Illuminant	CIE 1976 (u'_0, v'_0) chromaticity coordinates		CIE 1976 $((u'_0)_{10}, (v'_0)_{10})$ chromaticity coordinates	
	u'_0	v'_0	$(u'_0)_{10}$	$(v'_0)_{10}$
CIE ILL C	0.2009	0.4609	0.2000	0.4626
CIE ILL D$_{65}$	0.1978	0.4683	0.1979	0.4695
CIE ILL A	0.2560	0.5243	0.2590	0.5242
E	0.2105	0.4737	0.2105	0.4737

7.5 Calculation of the *permanency rating* of artists' pigments. Permanency rating $R = 10 - 10\,\Delta E_T/\Delta E_R$, where ΔE is the CIELAB (1976) [or ANLAB (40)] color difference that results over the test period during which a color sample is exposed to a standard illumination. Subscripts T and R refer to a test sample and the reference sample, respectively. The paint film of the reference sample contains madder lake, which is proposed as a pigment of marginal permanency in fine-arts applications; it provides the zero point of the permanency scale, whose range is 0–10 [Ref. 5.1, p. 48].

7.6 If the chromaticity coordinates (x, y) are given for the color of an object illuminated by CIE ILL A but with chromatic adaptation to CIE ILL C, the chromaticity coordinates (x', y') of *corresponding colors* (Sect. 7.12) with adaptation to CIE ILL A can be calculated by the use of the following equations from [Ref. 1.18, p. 361]:

$$x' = \frac{0.681x - 0.931y + 0.473}{-2.646x - 3.258y + 3.800} \; ,$$

$$y' = \frac{y}{-2.646x - 3.258y + 3.800} \; .$$

7.7 The *general color-rendering index* R_a is determined by use of

$$R_a = 100 - 4.6\Delta E_a \; ,$$

where ΔE_a represents the average value of the color difference ΔE for the eight standard colors when illuminated by two different sources (test source and reference source). A *special color-rendering index R* is related to the average ΔE for a selected reference color, or for a selected group of colors, by the same function.

The CIE recommends that the CIE 1964 (u, v) color-difference formula (not discussed in this book) be used in the calculation of ΔE. However the use of the CIELUV (1976) color-difference equation would give substantially the same color-rendering indexes [Ref. 4.4, p. 216]. A color difference determined from the CIELAB (1976) color-difference equation leads to a significantly different value of R_a, but there is evidence that the color-rendering index thus obtained correlates better with subjective observations [7.48]. Chromatic adaptation should be considered in the above calculation if the difference between the test and reference illumination sources cannot be neglected [7.47].

8.1 The Munsell color samples are sometimes employed to represent colors of perceptually equal spacing. The Munsell system was not determined exclusively by visual judgments and, as a result, it is very questionable whether the words "perceptually equal" can generally be used to describe the Hue and Chroma scales. The steps of Hue are not equal at all Hues, for constant Value and Chroma. Scales of Chroma, for constant Value, are not equal for various Hues. Even at constant Hue, a single Chroma scale does not usually have perceptually equal steps. The Value scale, however, was based on visual judgments and is uniform. By contrast, the OSA-UCS system (Chap. 9), developed decades later, was determined exclusively by visual judgments of relative sizes of perceptual differences of pairs of colors well distributed over the whole domain of attainable (permanent) colors.

8.2 *Darkness degree D* may be determined by the use of either [Ref. 8.13, Fig. 3] or

$$D = 10 - 6.1723 \log(40.7 \, h + 1) \ ,$$

where h is the relative luminance. Relative luminance is given by $h = Y/Y_m$, where Y is the luminance factor of the surface and Y_m is the maximum luminance factor of a color of the same chromaticity determined from a table giving the MacAdam limits (CIE ILL C), which is available in DIN-6164 and [Refs. 3.14, Table 6.11; 7.5].

8.3 Commenting on the judgement of NCS chromaticness C, Wright points out that the observer must have some concept of the appearance of the pure hues of colors at maximum saturation and asks, "Yet how can an observer acquire the ability to make this kind of judgement unless he first carries out some adaptation experiments to determine the maximum perceptible saturation of the different hues?" [3.12, see also 8.26].

8.4 The relation between the NCS and the Munsell color system has recently been studied. A report on that study has been prepared by A. Bencuya and F.W. Billmeyer, Jr.; it will probably be published in 1987 in *Color Research and Application.*

9.1 The OSA $(L, j, g,)$ notation is concise and logical; it is readily applied. But, for general use of the OSA collection of color samples by artists and designers, especially those without a technical background, there is a need for a simpler notation, equally concise and logical. The avoidance of negative numbers by the use of axes labeled j, g, b, and r is mentioned in Sect. 9.2. A less cumbersome scheme would retain the j and g axes but replace the numbers for the coordinates at the central point (0,0,0) by (20,20,20). The unit of 2 representing the distance between any two nearest-neighbor points would remain unchanged. Values of L, j, and g in the OSA system would be converted to the proposed scheme simply by adding 20 to each value. Thus, OSA designation $(-2,2,-4)$ would become (18,22,16).

For general use, the identification of the seven series of horizontal, vertical, and oblique cleavage planes by letters and numbers has been adopted in this book in place of the equations (representing lines plotted in Figs. 9.12, 13, 17–20 for the vertical and oblique planes) suggested by the OSA committee. Table 12.5 compares the OSA notations for the planes with those adopted in this book and with those proposed in which negative numbers are avoided.

9.2 Figure 9.8 presents radial lines that divide all horizontal planes in OSA color space into hue domains. The lines were drawn so that all points that represent actual OSA colors projected onto one horizontal plane were divided into groups characterized by the ten major Munsell Hues. Only points that have designations within two Munsell Hue units on either side of a radial line were taken into consideration (for example, within the range of 2.0PB to 8.0B). Because, for a given OSA chromaticness (j,g), Munsell Hue is not generally constant from one lightness level to the next, perfect segregation of the hue zones by one set of radial lines cannot be expected. The lines drawn appear to give the best fit to the Munsell Hue data (230 near-the-border colors) [8.29,30]. Minor border crossings to the "wrong" side were observed for 21 colors. Table 12.6 lists the slopes of the radial hue lines and the designations of the colors of the samples whose points fall on the "wrong" side of the radial lines.

9.3 When the OSA system was being designed, D.B. Judd specified the direction for the j axis so as to make it coincide with 10Y of the Munsell

Table 12.5. Comparison of notations for cleavage planes in OSA color space

		OSA notation	Notation used in this book	Notation without negative numbers
Horizontal cleavage planes		$L = 1$	$L(1)$	$L(21)$
		$L = 0$	$L(0)$	$L(20)$
		$L = -1$	$L(-1)$	$L(19)$
Vertical cleavage planes	E series	$j - g = 2$	$E(2)$	$E(22)$
		$j - g = 0$	$E(0)$	$E(20)$
		$j - g = -2$	$E(-2)$	$E(18)$
	F series	$j + g = 2$	$F(2)$	$F(22)$
		$j + g = 0$	$F(0)$	$F(20)$
		$j + g = -2$	$F(-2)$	$F(18)$
Oblique cleavage planes	J series	$L + j = 2$	$J(2)$	$J(22)$
		$L + j = 0$	$J(0)$	$J(20)$
		$L + j = -2$	$J(-2)$	$J(18)$
	B series	$L - j = 2$	$B(2)$	$B(22)$
		$L - j = 0$	$B(0)$	$B(20)$
		$L - j = -2$	$B(-2)$	$B(18)$
	G series	$L + g = 2$	$G(2)$	$G(22)$
		$L + g = 0$	$G(0)$	$G(20)$
		$L + g = -2$	$G(-2)$	$G(18)$
	R series	$L - g = 2$	$R(2)$	$R(22)$
		$L - g = 0$	$R(0)$	$R(20)$
		$L - g = -2$	$R(-2)$	$R(18)$

system (the hue radius separating the GY and Y hue ranges). Figure 9.8 shows the radial hue-division line separating OSA GY and Y samples to deviate somewhat above (slope 2/21) the j axis.

9.4 For consistency with Figs. 9.1, 9.5 and 9.22, the L scale shown in Figs. 9.17 – 9.20, 9.28 and 9.29 would need to be expanded by $\sqrt{2}$. The resulting inclination of an "edge-on" profile would be 54.7° rather than 45°.

10.1 The abbreviations shown in Fig. 10.2 are those originally proposed in the ISCC-NBS scheme [Ref. 7.2, Fig. 11] except for "viv." for "vivid" (originally "v.") and "dp." for "deep" (no abbreviation for "deep" was given in the original scheme). These two changes appear in [Ref. 1.38, Fig. 1, p. 448].

11.1 The Munsell Hue equivalents of the NCS unitary hues and their intermediate hues (Table 11.1) were estimated by superposing plots of NCS hue lines and Munsell Hue lines on a CIE(u', v') chromaticity diagram [Ref. 11.1, Figs. 16,18]. The Munsell Hue numbers for the unitary hues may be compared with those based on data tabulated by Kuehni [Ref. 1.39, p. 60]: red,

Table 12.6. Specifications of the radial hue lines that divide all horizontal planes in OSA color space into hue domains and of the 21 sample colors (OSA-UCS) that do not fit (Fig. 9.8)

Line no.	Pair of Munsell Hue zones separated by the radial hue line	Slope of the radial hue line	Designations of colors that are on the "wrong" side of the radial hue line	
			Munsell Hue	OSA (L, j, g)
1	Y GY	2/21	0.4GY	$(-4.0, \quad 4.0, \quad 0.3)$
2	GY G	18/23	none	
3	G BG	4	9.4G	$(-3.1, \quad 1.2, \quad 4.9)$
			8.9G	$(-5.1, \quad 0.8, \quad 3.3)$
			0.7BG	$(1.5, \quad 0.4, \quad 1.5)$
			0.3BG	$(-0.4, \quad 0.4, \quad 1.5)$
4	BG B	−16/5	9.7BG	$(-0.5, \quad -0.5, \quad 1.5)$
			0.3B	$(1.5, \quad -0.4, \quad 1.4)$
5	B PB	− 9/10	9.8B	$(-4.0, \quad -2.1, \quad 1.8)$
			9.7B	$(-6.9, \quad -1.2, \quad 0.9)$
			8.8B	$(-5.1, \quad -1.0, \quad 0.8)$
			0.3PB	$(4.2, \quad -1.9, \quad 1.9)$
			0.3PB	$(1.9, \quad -4.0, \quad 3.8)$
6	PB P	1/10	0.5P	$(0.9, \quad -1.1, \quad -0.1)$
7	P RP	16/13	none	
8	RP R	−18	8.4RP	$(2.2, \quad 0.2, \quad -2.2)$
			0.8R	$(-5.6, \quad 0.1, \quad -5.7)$
			1.0R	$(-6.0, \quad 0.1, \quad -4.1)$
9	R YR	−10/7	0.1YR	$(-4.7, \quad 2.7, \quad -4.6)$
			0.6YR	$(-3.9, \quad 3.8, \quad -5.6)$
10	YR Y	− 7/17	9.4YR	$(1.9, \quad 10.1, \quad -3.9)$
			0.2Y	$(-1.0, \quad 1.9, \quad -0.9)$
			0.5Y	$(1.1, \quad 2.8, \quad -1.2)$

5; yellow, 28; green, 46; blue, 74. It must be recognized that the NCS unitary hues are averages established for selected background (white), illumination (daylight), and adaptation (neutral). The Munsell Hue equivalents for the unitary hues given in Table 11.1 are approximations because, as G. Tonnquist has reported (his information is tabulated in [8.26]), the equivalent Munsell Hue may vary when there are changes of Chroma or Value.

11.2 If the CIE(x, y, Y) notation is given for a color, the HBS hue and saturation coefficients can be calculated by use of equations presented in [11.2].

11.3 Professor Hurvich has brought to my attention the dissertation by J.M. Eichengreen [12.5], in which the variability of afterimages and their complex dependency on experimental conditions are discussed. The experiments of Wilson and Brocklebank are recognized as thorough [11.1], but application of their data to conditions that imply viewing conditions that differ significantly from theirs may not be justified. Their recommendation that afterimage complementary color pairs (of the type they observed) be used to design a color circle for artistic applications (e.g., painting) appears to be amply justified.

11.4 In testing for the presence of induced color in colored shadows, it is advisable to avoid the possible occurrence of other effects, by viewing the shadow area through the black tube before introducing the induction stimulus (e.g., the red light that induces green). After, for example, the red beam is projected, the induced color is not seen until the black tube is allowed to drop from the eye [Refs. 7.43; 11.14, p. 272].

12.1 Table 12.1 and a set of nine Munsell-CIE (1931) charts were published originally in an OSA report [8.20]. The charts were prepared in the office of D. Nickerson at the U.S. Department of Agriculture, Washington, D.C., to represent the central most-used portion of the original full-scale charts that had been prepared by Nickerson and S.M. Newhall for the OSA report. In 1964, new charts were prepared in Miss Nickerson's laboratory at the request of the Munsell Color Company [now, Munsell Color (Sect. 12.1)], Baltimore, Md., that would extend to slightly different and perhaps more useful x, y limits. These 1964 charts, prepared from the original full-scale Newhall-Nickerson charts, were supplied to the Munsell Color Company, and they are currently being published by them in large format. Permission to reproduce the charts has been granted by Munsell Color.

Figures 12.1–9 (in small format) show tracings of the curves with minor changes in presentation. The indication in full of the Hue designation near the end of each Hue line and the inclusion of Munsell Hue numbers are modifications that have been introduced. If greater accuracy is desired, the reproductions of the original graphs (showing centimeter and millimeter grids) in large format (full size) in the Munsell edition [8.44] or in small format in [3.14,8.19a] should be consulted.

Plate I. *Top*: A six-member color circle. Opposing additive complementary color pairs (Sect. 7.2). Hues (Munsell notation): 5G, 5.5Y, 2.5R, 5.5RP, 3PB, 4.5BG (Fig. 7.2). *Bottom:* A Goethe color circle. Opposing afterimage complementary color pairs (Sect. 11.7). Hues (approximate Munsell notation): 5G, 8.5Y, 10R, 5.5RP, 10PB, 10B. The color samples were selected with the aid of Fig. 11.7. The samples were photographed; hence the reproduction of their colors here must be considered as only approximate (See p. 45)

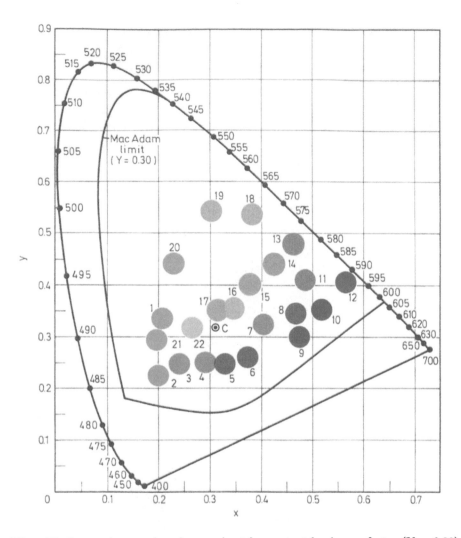

Plate II. Some color samples of approximately constant luminance factor ($Y = 0.30$) (Table 7.2). The chromaticities of the colors of the samples are compared with the MacAdam limit ($Y = 0.30$). The samples were cut from standard glossy color chips (Munsell) and from glossy chips in [7.7]. The samples were photographed; hence the reproduction of their colors here must be considered as only approximate. CIE 1931 (x, y) chromaticity diagram. See p. 76 and Plate VII

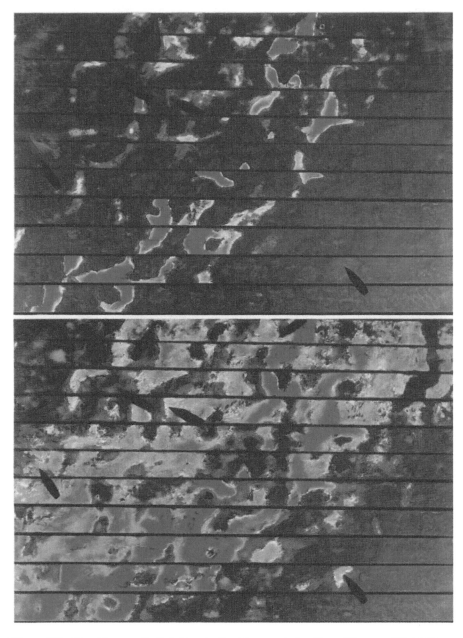

Plate III. "Les Bateaux". Painting in liquid crystals by Yves Charnay (50 × 65 cm, 1980). The colors seen vary with ambient temperature: (*top*) 18°C, (*bottom*) 23°C. (Courtesy of the artist.) See p. 83

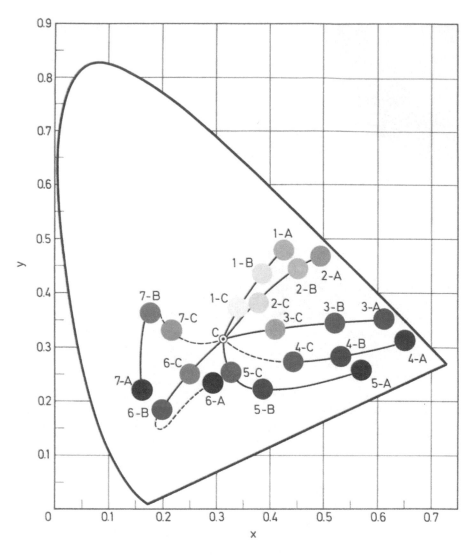

Plate IV. Chromaticities of the colors of glossy paint film samples. Seven pigments and two mixtures of each of them with titanium dioxide white pigment are represented. The curves (mixture lines) show the variation of chromaticity with white pigment content (Table 7.3). Luminance factor range: $Y = 0.004$ to 0.824, CIE ILL C. The samples were cut from color chips in [7.7]. The samples were photographed; hence the reproduction of their colors here must be considered as only approximate. See p. 84

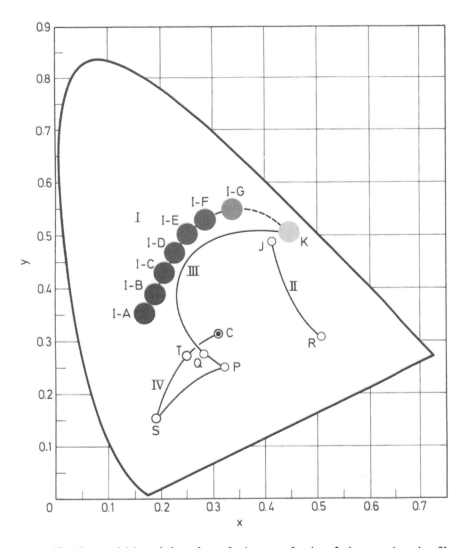

Plate V. Chromaticities of the colors of mixtures of pairs of pigments in paint films. Mixture lines are shown for: (*I*) Milori blue and chrome yellow (*K*) (Table 7.4); (*II*) zinc yellow (*J*) and deep cadmium red (*R*) (reproduced from [Ref. 2.1, Fig. 18.7]); (*III*) Prussian blue (*P*) and chrome yellow (*K*) in oil (reproduced from [Ref. 7.21, Fig. 12]); and (*IV*) Prussian blue (*P*) and lead white (*C*) in oil (reproduced from [Ref. 7.21, Fig. 7]). CIE ILL C. The samples were cut from glossy color chips in [7.7]. The samples were photographed; hence the reproduction of their colors here must be considered as only approximate. See p. 85

Plate VI. Undesired grayness in pointillism. *Top:* If the mosaic pattern of blue and yellow is viewed from a sufficient distance, the pattern fuses, producing a grayish yellow. *Bottom:* A mosaic pattern of the bluish green and purplish red colors would fuse and produce a neutral gray. The display of glossy color samples was photographed; hence the reproduction of their colors here must be considered as only approximate. See p. 88

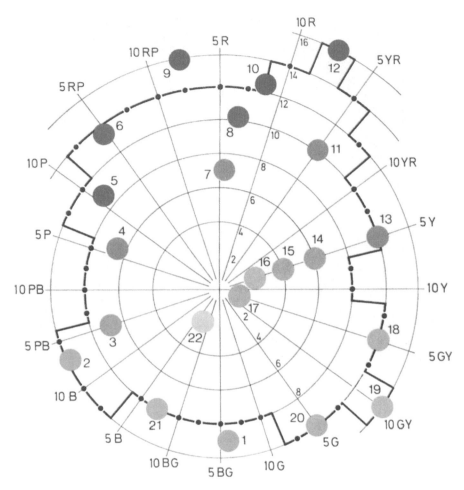

Plate VII. Color samples (glossy) shown on a Munsell Hue-Chroma diagram. Some of the samples were cut from standard Munsell chips (Value 6); the others were cut from commercial color chips (approximately Value 6). The colors (numbered) are the same as those displayed in Plate II and described in Table 7.2. The black dots, which represent Munsell samples of maximum Chroma, have been transferred from Fig. 8.10. The heavy black line delineates the outer Chroma limit of available glossy Munsell samples. The samples were photographed; hence the reproduction of their colors here must be considered as only approximate. See p. 117

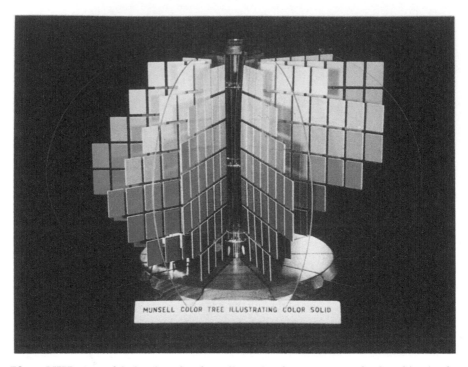

Plate VIII. A model showing the three-dimensional arrangment of color chips in the Munsell color solid. There are 40 basic radial Hue planes in the Munsell system; 10 are included in this model. (Photograph courtesy of Munsell Color, Baltimore.) See p. 120

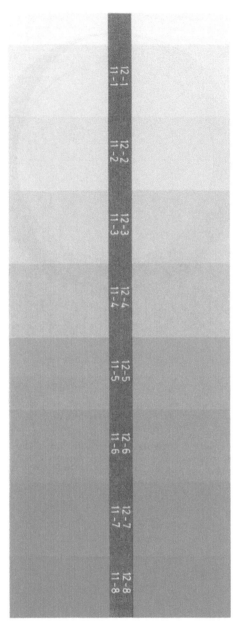

12-1 / 11-1
12-2 / 11-2
12-3 / 11-3
12-4 / 11-4
12-5 / 11-5
12-6 / 11-6
12-7 / 11-7
12-8 / 11-8

Plate IX. Reproduction of one card from the *Munsell Limit Color Cascade* showing colors varying in Cascade lightness level from 1 to 8 in two Hue series, 11 and 12 (maximum available Chroma) (photoreduction to 60 % of full size). Although each color step is of uniform lightness, each is perceived to vary in lightness [the Mach- band effect (Sect. 11.10)]. The visual phenomenon disappears if a black common boundary line is drawn between adjacent steps. The card was photographed; hence the reproduction of the colors here must be considered as only approximate. (Reproduced with the permission of Munsell Color, Baltimore.) See p. 123

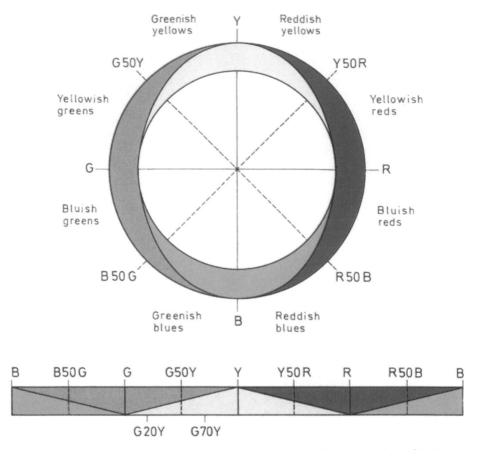

Plate X. Hering hue circle. A schematic representation of the unitary hues (Y, R, B, and G) and the binary compositions of all other hues. (The hues of the printing inks employed here are not the unitary hues.) (Based on [Ref. 8.22, Fig. 8].) See p. 134

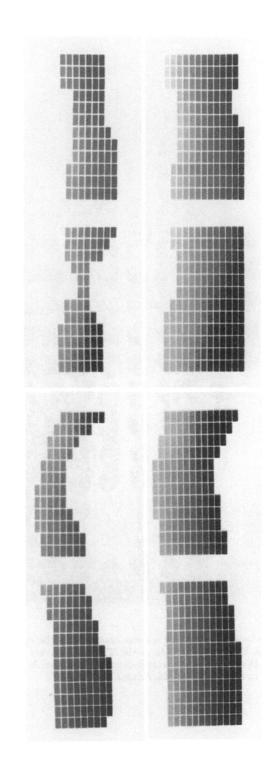

Plate XI. Chroma Cosmos 5000. Four folded charts of constant Munsell Chroma showing sample arrays varying (vertically) in Munsell Value and (horizontally) in Munsell Hue (Fig. 8.26). The upper charts represent Munsell Chroma 9 and the lower ones Munsell Chroma 6. (Portion of photograph. Courtesy of T. Hosono, Japan Color Research Institute, Tokyo). See p. 138

Plate XII. Model showing 424 spheres coated with the same paints that were used to coat the OSA-UCS color sample cards. The spheres are mounted at specified points to illustrate the geometrical arrangement of the sample colors in OSA color space. (Photograph courtesy of D.L. MacAdam). See p. 143

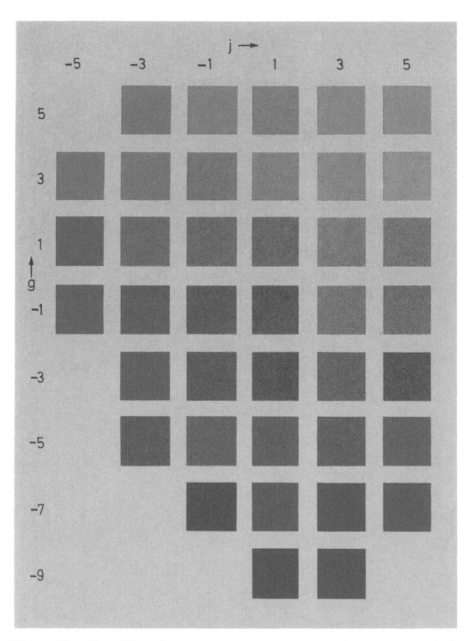

Plate XIII. The OSA uniform color array in horizontal cleavage plane $L(-3)$, representing constant OSA lightness $L = -3$. (The spheres representing the color array in this horizontal cleavage plane can be located in Plate XII.) The differences between any color and its four nearest neighbors are identical. The color squares were cut from color cards of an OSA-UCS set; their colors are identified by notations obtainable from Fig. 9.9. The array of color squares was photographed; hence the reproduction of their colors here must be considered as only approximate. See p. 150

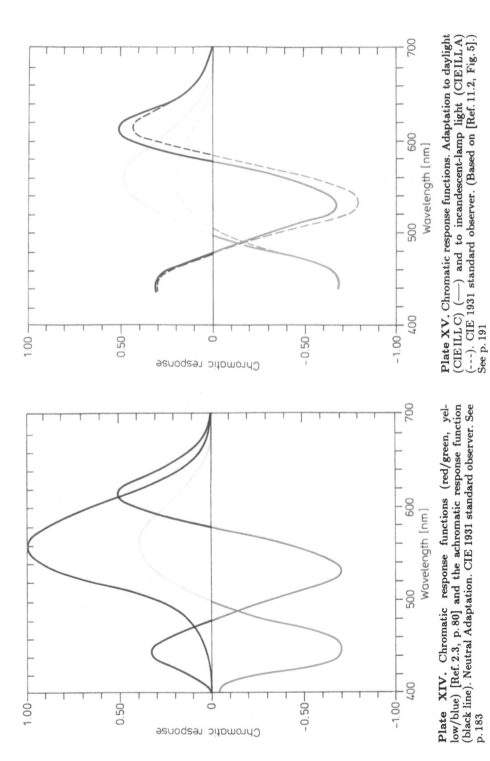

Plate XIV. Chromatic response functions (red/green, yellow/blue) [Ref. 2.3, p. 80] and the achromatic response function (black line). Neutral Adaptation. CIE 1931 standard observer. See p. 183

Plate XV. Chromatic response functions. Adaptation to daylight (CIE ILL C) (——) and to incandescent-lamp light (CIE ILL A) (---). CIE 1931 standard observer. (Based on [Ref. 11.2, Fig. 5].) See p. 191

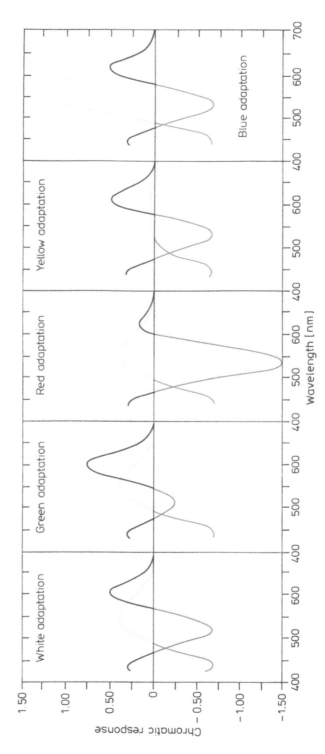

Plate XVI. Chromatic response functions. Adaptation to white light and to light that produces unitary hues [monochromatic light of wavelength 475 nm (blue), 500 nm (green), 580 nm (yellow), and red light produced by a mixture, 670 and 440 nm]. CIE 1931 standard observer. (Based on [Ref. 11.5, Figs. 1–5].) See p. 191

263

Plate XVIII. Demonstration of an afterimage viewed against a yellow background. See Plate XVII. The afterimage of the square is orange. See p. 195

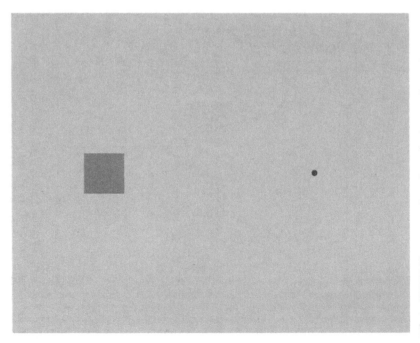

Plate XVII. Demonstration of an afterimage of complementary hue. Gaze steadily at the green square for 20 seconds. Then focus the eyes on the dot in the area below. The afterimage of the square is magenta. See p. 195

Plate XX. Demonstration of the afterimage of a neutral gray square (having green surrounds) against a neutral gray background. See Plate XVII. The afterimage of the square is green. See pp. 195,199

Plate XIX. Demonstration of an afterimage viewed against a red background. See Plate XVII. The afterimage of the square is red, of greater vividness than the red of the background. See p. 195

265

Plate XXI. "Glowing" diagonals (heightened contrast) in a pattern. A manifestation of edge contrast. See p. 208

Plate XXII. "Dimmed" diagonals (heightened contrast) in a pattern. A manifestation of edge contrast. See p. 208

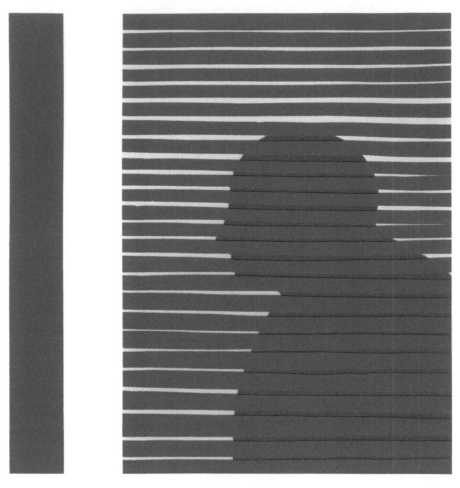

Plate XXIII. Demonstration of assimilation. Illustration made from three uniformly colored papers. The red strips on the yellow background appear yellowish; the red strips on the blue background appear bluish. The rectangular area on the left, cut from the same red paper, is included for comparison. See p. 208

References

Chapter 1

1.1   J.W. von Goethe: *Farbenlehre* (Color Theory) (1810). C.L. Eastlake's Translation (1840). Introduction by D.B. Judd (M.I.T.Press, Cambridge, Mass. 1970)
1.2   R. Matthaei (ed.): *Goethe's Color Theory*. Translated by H. Aach (Van Nostrand Reinhold, New York 1970)
1.3   F. Birren: A sense of illumination. Color Res. Appl. **2** (2), 69–74 (1977)
1.4   A. Garrett: Report on *Color 77*, Third Congress of the International Color Association. Leonardo **11**, (1), 41–42 (1978)
1.5   F. Birren: Color perception in art: Beyond the eye into the brain. Leonardo **9**, (2), 105–110 (1976)
1.6   M.E. Chevreul: *The Principles of Harmony and Contrast of Colors and Their Applications in the Arts* (1839). Reprinted. Introduction and Notes by F. Birren (Van Nostrand Reinhold, New York 1967)
1.7   O.N. Rood: *Modern Chromatics: Students' Text-Book of Color with Applications to Art and Industry* (1879). Reprinted. Introduction and Notes by F. Birren (Van Nostrand Reinhold, New York 1973)
1.8   J. Albers: *Interaction of Color* (Yale University Press, New Haven, Conn. 1971)
1.9   J.H. Holloway, J.A. Weil: A conversation with Josef Albers. Leonardo **3**, (4), 459–464 (1970)
1.10   *J. Albers: His Work as Contribution to Visual Articulation in the Twentieth Century* (George Wittenborn, New York)
1.11   D. Nickerson: History of the Munsell color system and its scientific application. J. Opt. Soc. Am. **30**, 575–586 (1940)
1.12   D. Nickerson: History of the Munsell color system. Color Eng. **7** (5), 42–51 (1969)
1.13   W. Faulkner: *Architecture and Color* (Wiley, New York 1972)
1.14   F. Birren (ed.): *A Grammar of Color: A Basic Treatise on the Color System of Albert H. Munsell* (Van Nostrand Reinhold, New York 1969)
1.15   W. Ostwald: *Color Science: A Handbook for Advanced Students in Schools, Colleges, and the Various Arts, Crafts, and Industries Depending on the Use of Color.* Translated by J. Scott Taylor (Winsor and Newton, London 1931)
1.16   F. Birren (ed.) *Ostwald: The Color Primer* (Van Nostrand Reinhold, New York 1969)
1.17   E. Jacobson: *Basic Color: An Interpretation of the Ostwald Color System* (Paul Theobald, Chicago 1948)
1.18   D.B. Judd, G. Wyszecki: *Color in Business, Science and Industry*, 3rd ed. (Wiley, New York 1975)
1.19   D.B. Judd: *Color in Our Daily Lives*, NBS Consumer Information Series 6 (U.S. Dept. of Commerce, Washington, D.C. 1975)
1.20   R.W. Burnham, R.M. Hanes, C.J. Bartleson: *Color: A Guide to Basic Facts and Concepts* (Wiley, New York 1963)
1.21   J.M. Carpenter: *Color in Art: A Tribute to Arthur Pope* (Fogg Art Museum, Harvard University, Cambridge, Mass. 1974)
1.22   A. Pope: *The Language of Drawing and Painting* (Harvard University Press, Cambridge, Mass. 1949).(Russell and Russell, New York 1967)
1.23   R.B. Farnum: Results of a questionnaire on color in art education. J. Opt. Soc. Am. **32**, 720–726 (1942)

1.24 J.T. Luke: Toward a new viewpoint for the artist. Color Res. Appl. **1** (1), 23–36 (1976)

1.25 J.T. Luke: Simulation of colored illumination. Color Res. Appl. **9** (3), 153–156 (1984)

1.26 G. Marcus: A color system for artists. Leonardo **9** (1), 48–51 (1976)

1.27 M.L. Meixner: Instruction on light and color in art at the Iowa State University. Leonardo **9** (1), 52–55 (1976)

1.28 L. Swirnoff: Experiments on the interaction of color and form. Leonardo **9** (3), 191 -195 (1976)

1.29 H. Thomas: Application of the Ostwald Color System in my painting. Leonardo **13** (1), 11–16 (1980)

1.30 H. Thomas: Simulation of colored illumination in painting using the Ostwald system. Color Res. Appl. **9** (3), 147–152 (1984)

1.31 J.C. Fish: Color as sensation in visual art and in science. Leonardo **14** (2), 89–98 (1981)

1.32 E. Jacobson, W.C. Granville, C.E. Foss: *Color Harmony Manual*, 3rd ed. (Container Corporation of America, Chicago 1948)

1.33 *SIS Colour Atlas NCS (Natural Colour System)* Swedish Standard No. 01 91 02 (1979) (Scandinavian Colour Institute, Stockholm)

1.34 *Munsell Book of Color.* Glossy Finish Collection and Matt Finish Collection (Munsell Color, Macbeth Division of Kollmorgen Corporation, Baltimore, Md. 1976)

1.35 *OSA Uniform Color Scales* (Optical Society of America, Washington, D.C. 1977)

1.36 *Chroma Cosmos 5000* (Japan Color Research Institute, Tokyo 1978)

1.37 *Chromaton 707* (Japan Color Research Institute, Tokyo 1982)

1.38 *Webster's Third New International Dictionary* (Unabridged) (G. and C. Merriam Co., Springfield, Mass. 1971)

1.39 R.G. Kuehni: *Color, Essence and Logic* (Van Nostrand Reinhold, New York 1983)

Chapter 2

2.1 R.M. Evans: *An Introduction to Color* (Wiley, New York 1948)

2.2 C.A. Padgham, J.E. Saunders: *The Perception of Light and Colour* (G. Bell, London 1975)

2.3 L.M. Hurvich: *Color Vision* (Sinauer Associates, Sunderland, Mass. 1981)

2.4 C.F. Bohren, A.B. Fraser: Colors of the sky. Phys. Teach. **23** (5), 267–272 (1985)

2.5 R.M. Evans: *The Perception of Color* (Wiley, New York 1974)

2.6 D. Jameson, L.M. Hurvich: From contrast to assimilation: In art and in the eye. Leonardo **8** (2), 125–131 (1975)

2.7 C.J. Bartleson: On chromatic adaptation and persistence. Color Res. Appl. **6** (3), 153–160 (1981)

2.8 D. Jameson, L.M. Hurvich: "Color Adaptation: Sensitivity, Contrast, After-Images", in *Handbook of Sensory Physiology*, Vol. VII/4 *Visual Psychophysics*, ed. by D. Jameson, L.M. Hurvich (Springer, Berlin, Heidelberg 1972)

Chapter 3

3.1 *International Lighting Vocabulary.* CIE Publication No. 17 (E-1.1) (CIE, Paris 1970)

3.2 R.M. Johnston, R.E. Park: Color and appearance. Color Eng. **4** (6), 14–19 (1976)

3.3 R.W.G. Hunt: The specification of colour appearance. I. Concepts and terms. Color Res. Appl. **2** (2), 55–68 (1977)

3.4 K. Richter: Cube-root color spaces and chromatic adaptation. Color Res. Appl. **5** (1), 25–43 (1980)

3.5 A.R. Barlee: Uniform color spaces and colorimeter performance. J. Oil Colour Chem. Assoc. **49** (4), 275–298 (1966)

3.6   D. Jameson, L.M. Hurvich: The science of color appearance. Color Eng. **5** (5), 29–36, 43 (1967)

3.7   L.M. Hurvich, D. Jameson: Some quantitative aspects of an opponent-colors theory. II. Brightness, saturation, and hue in normal and dichromatic vision. J. Opt. Soc. Am. **45**, 602–616 (1955)

3.8   J.C. Fish: *The Perception of Color* by Ralph M. Evans (Book review). Color Res. Appl. **2** (4), 197–199 (1977)

3.9   R.M. Evans: Fluorescence and its appearance. J. Color Appearance **1** (4), 4 (1972)

3.10   R.M. Evans: "The Perception of Color", in *Advances in Chemistry Series 107* (American Chemical Society, Washington, D.C. 1971) pp. 43–68

3.11   D.B. Judd: *The Language of Drawing and Painting* by Arthur Pope (Book review). J. Opt. Soc. Am. **40**, 122 (1950)

3.12   W.D. Wright: The basic concepts and attributes of colour order systems. Color Res. Appl. **9** (4), 229–233 (1984)

3.13   A.R. Robertson: Colour order systems: An introductory review. Color Res. Appl. **9** (4), 234–240 (1984)

3.14   G. Wyszecki, W.S. Stiles: *Color Science: Concepts and Methods, Quantitative Data and Formulas* (Wiley, New York 1967) (2nd ed. 1982)

3.15   H. Osborne (ed.): *The Oxford Companion to Art* (Oxford University Press, Oxford 1970)

Chapter 4

4.1   K.L. Kelly: Color designations for lights. J. Opt. Soc. Am. **33**, 627–632 (1943)

4.2   P.K. Kaiser, J.P. Comerford, D.M. Bodinger: Saturation of spectral lights. J. Opt. Soc. Am. **66**, 818–826 (1976)

4.3   K. Uchikawa, H. Uchikawa, P.K. Kaiser: Equating colors for saturation and brightness: The relationship to luminance. J. Opt. Soc. Am. **72**, 1219–1224 (1982)

4.4   D.L. MacAdam: *Color Measurement, Theme and Variations*, Springer Ser. Opt. Sci., Vol. 27 (Springer, Berlin, Heidelberg 1981) (2nd ed. 1985)

4.5   "Laser and Maser", in *The New Encyclopaedia Britannica*, 15th ed., Vol. 10 (1984) pp. 686–689

4.6   S. Ostoja-Kotkowski: Audio-kinetic art with laser beams and electronic systems. Leonardo **8** (2), 142–144 (1975)

4.7   S. Ostoja-Kotkowski: Audo-kinetic art: The construction and operation of my "laser-chromasonic tower". Leonardo **10** (1), 51–53 (1977)

4.8   C.L. Strong: How to construct an argon gas laser with outputs at several wavelengths. Sci. Am. **220** (2), 118–123 (1969)

4.9   P. Sorokin: Organic lasers. Sci. Am. **220** (2), 30–40 (1969)

4.10   F.P. Schäfer (ed.): *Dye Lasers*, 2nd ed., Topics Appl. Phys. Vol. 1 (Springer, Berlin, Heidelberg 1977)

4.11   F.C. Strome: The dye laser. Eastman Org. Chem. Bull. **46** (2), 1–4 (1974)

4.12   C.L. Strong: A tunable laser using organic dye is made at home for less than \$75. Sci. Am. **222** (2), 116–120 (1970)

4.13   M. Bass, T.F. Deutsch, M.J. Weber: "Dye Lasers", in *Lasers*, ed. by A.K. Levine, A.J. De Maria, Vol. II (Marcel Dekker, New York 1971) p. 270

4.14   T. Kallard: *Exploring Lase Light* (Optosonic Press, New York 1977)

4.15   F.G. McAleese: *The Laser Experimenter's Handbook* (Tab Books, Blue Ridge Summit, Penn. 1979)

4.16   N. Ohta, G. Wyszecki: Colorimetric significance of mercury-emission lines in fluorescent lamps. J. Opt. Soc. Am. **65**, 1354–1358 (1975)

4.17   A.H. Taylor, G.P. Kerr: The distribution of energy in the visible spectrum of daylight. J. Opt. Soc. Am. **31**, 3–8 (1941)

4.18   OSA (Committee on Colorimetry, Optical Society of America): *The Science of Color* (T.Y. Crowell, New York 1953) (Optical Society of America, Washington, D.C. 1963)

4.19   G. Wyszecki: Development of new CIE standard sources for colorimetry. Die Farbe **19** (1–6), 43–76 (1970)

Chapter 5

5.1   H.W. Levison: *Artists' Pigments: Lightfastness Tests and Ratings* (Colorlab, Hallandale, Florida 1976)
5.2   R. Mayer: *The Artist's Handbook of Materials and Techniques*, 3rd ed. (Thomas Nelson, London 1975)
5.3   K. Wehlte: *The Materials and Techniques of Painting*, translated by Ursus Dix (Van Nostrand Reinhold, New York 1975)
5.4   R.J. Gettens, G.L. Stout: *Painting Materials: A Short Encyclopaedia* (Dover, New York 1966)
5.5   H.W. Levison: "Pigmentation of Artists' Colors", in *Pigment Handbook*, Vol. III, ed. by T.C. Patton (Wiley, New York 1973) pp. 423–434
5.6   *Colour Index*, 3rd ed., 2nd revision (American Association of Textile Chemists and Colorists, Research Triangle Park, N.C.; and Society of Dyers and Colourists, Bradford, England 1982)
5.7   E. Wich: The *Colour Index*. Color Res. Appl. **2** (2), 77–80 (1977)
5.8   F.W. Billmeyer, Jr., M. Saltzman: *Principles of Color Technology* (Wiley, New York 1966) (2nd ed. 1981)
5.9   N.F. Barnes: A spectrophotometric study of artists' pigments. Tech. Stud. Field Fine Arts **7** (3), 120–138 (1935)
5.10 N.F. Barnes: Color characteristics of artists' pigments. J. Opt. Soc. Am. **29** (5), 208–214 (1939)
5.11 H.R. Davidson: Formulations of the OSA Uniform Color Scales Committee samples. Color Res. Appl. **6** (1), 38–52 (1981)
5.12 R. Bowman: Paintings with fluorescent pigments of the microcosm and macrocosm. Leonardo **6** (4), 289–292 (1973)
5.13 A.K. Schein, W.R. Dana: Fluorescent pigments. Paint Varn. Prod. **60** (8), 72–76 (1970)
5.14 R.W. Voedish, D.W. Ellis: "Fluorescent Pigments (Daylight)", in *Kirk-Othmer Encyclopedia of Chemical Technology*, 2nd ed., Vol. IX (Wiley, New York 1966) pp. 483–506
5.15 R. Donaldson: Spectrophotometry of fluorescent pigments. Br.J. Appl. Phys. **5**, 210–124 (1954)
5.16 G. Wyszecki: "Current Developments in Colorimetry", in *Colour 73*. Second Congress of the International Colour Association, York, England (Adam Hilger, London 1973) pp. 21–51
5.17 D. Jameson: Some misunderstandings about color perception, color mixture and color measurement. Leonardo. **16** (1), 41–42 (1983)
5.18 G.S. Wasserman: *Color Vision. An Historical Introduction* (Wiley, New York 1978)
5.19 F. Gerritsen: "Colour Teaching: A New Colour Circle", in *Colour 73* Second Congress of the International Colour Association, York, England (Adam Hilger, London 1973) pp. 494–498
5.20 F. Gerritsen: Evolution of the color diagram. Color Res. Appl. **4** (1), 33–38 (1979)
5.21 M.H. Wilson, R.W. Brocklebank: Complementary hues of after-images. J. Opt. Soc. Am. **45**, 293–299 (1955)
5.22 P.R. Hofstatter (ed.): *Gesetz im Grenzenlosen: Goethes naturwissenschaftliche Schriften* (Leykam-Verlag, Graz 1949)
5.23 C. Parkhurst, R.L. Feller: Who invented the color wheel? Color Res. Appl. **7** (3), 217–230 (1982)

Chapter 6

6.1   W.D. Wright: The golden jubilee of colour in the CIE 1931–1981. Color Res. Appl. **7** (1), 12–15 (1982)
6.2   W.D. Wright: *The Measurement of Colour*, 2nd ed. (Hilger & Watts, London 1958)
6.2a W.D. Wright: *The Measurement of Colour*, 4th ed. (Adam Hilger, London 1969)
6.3   D.L. MacAdam: Color measurement and tolerances. Off. Dig., Fed. Soc.Coatings Technol. **37** (491), 1488–1531 (1965)

6.4   S.A. Burns, V.J. Smith, J. Pokorny, A.E. Elsner: Brightness of equal-luminance lights. J. Opt. Soc. Am. **72** (9), 1225–1231 (1982)

6.5   S.B. Saunders, F.Grum: Measurement of luminance factor. Color Res. Appl. **2** (3), 121–123 (1977)

6.6   D.L. MacAdam: *Color Vision: An Historical Introduction* by G.S. Wasserman. (Book review) J. Opt. Soc. Am. **69** (4), 636–637 (1979)

6.7   D.B. Judd: "Basic Correlates of the Visual Stimulus", in *Handbook of Experimental Psychology*, ed. by S.S. Stevens (Wiley, New York 1951) Chap. 22

Chapter 7

7.1   D.L. MacAdam: On the geometry of color space. J. Franklin Inst. **238**, 195–210 (1944)

7.2   K.L. Kelly, D.B. Judd: *Color: Universal Language and Dictionary of Names*, NBS Special Publication 440 (U.S. Government Printing Office, Washington, D.C. 1976)

7.3   W.B. Warfel, W.R. Klappert: *Color Science for Lighting the Stage* (Yale University Press, New Haven, Conn. 1981)

7.4   D.L. McAdam: Theory of the maximum visual efficiency of colored materials. J. Opt. Soc. Am. **25**, 249–252 (1935)

7.5   D.L. MacAdam: Maximum visual efficiency of colored materials. J. Opt. Soc. Am. **25**, 361–367 (1935)

7.6   D.B. Judd, G. Wyszecki: Extension of the Munsell renotation system to very dark colors. J. Opt. Soc. Am. **46** (4), 281–284 (1956)

7.7   *Pigment Colors for Paint* (revised) (E.I. du Pont de Nemours Co., Wilmington, Del. 1957)

7.8   M.Saltzman: Colored organic pigments: Why so many? Why so few? Off. Dig., Fed. Soc. Coatings Technol. **35** (458), 245–257 (1963)

7.9   *Modular Colors* (Trade Booklet) (Permanent Pigments, Cincinnati, Ohio 1975)

7.10 *Enduring Colors for the Artist* (Permanent Pigments, Cincinnati, Ohio 1975)

7.11 W.D. Wright: Iridescence. Leonardo **7** (4), 325–328 (1974)

7.12 E. Verity: *Colour Observed* (Macmillan, London 1980)

7.13 L.M. Greenstein: "Pearlescence, the Optical Behavior of Nacreous and Interference Pigments", in *Pigment Handbook*, Vol, III, ed. by T.C. Patton (Wiley, New York 1973) pp. 357–390

7.14 T. Bryson: Creating interference colors on thermoplastic films without colorants. Color Res. Appl. **9** (1), 57–60 (1984)

7.15 R. Lemberg: Liquid crystals: A new material for artists. Leonardo **2** (1), 45–50 (1969)

7.16 D.M. Makow: Color properties of liquid crystals and their application to visual arts. Color Res. Appl. **4** (1), 25–32 (1979)

7.17 Y. Charnay: A new medium for expression: Painting with liquid crystals. Leonardo **15** (3), 219–221 (1982)

7.18 D.M. Makow: Liquid crystals in painting and sculpture. Leonardo **15** (4), 257–261 (1982)

7.19 K.L. Kelly, K.S. Gibson, D. Nickerson: Tristimulus specifications of the *Munsell Book of Color* from spectrophotometric measurements. J. Opt. Soc. Am. **33**, 355–376 (1943)

7.20 R.M. Johnston: "Color Theory", in *Pigment Handbook*, Vol. III, ed. by T.C. Patton (Wiley, New York 1973) pp. 229–288

7.21 S.R. Jones: "The History of the Artist's Palette in Terms of Chromaticity", in *Application of Science in Examination of Works of Art*. Proceedings of seminar (Museum of Fine Arts, Boston 1965) pp. 71–77

7.22 P.C. Goldmark, J.W. Christensen, J.J. Reeves: Color television – USA standard. Proc. IRE **39**, 1288–1313 (1951)

7.23 R.W.G. Hunt: *The Reproduction of Colour in Photography, Printing and Television*, 3rd ed. (Fountain Press, Kings Langley, England 1975)

7.24 M.R. Pointer: The Gamut of Real Surface Colors. Color Res. Appl. **5** (3), 145–155 (1980)

7.25 R.S. Hunter: Instrumental methods of color and color difference measurement. Am. Ceram. Soc., Bull. **36** (7), 249–255 (1957)

7.26 D.B. Judd: Ideal color space. Color Eng. **8** (2), 36–52 (1970)

7.27 S. Le Sota, et al. (compilers): *Paint/Coatings Dictionary* (Federation of Societies for Coatings Technology, Philadelphia 1978)

7.28 D.I. Morley, R. Munn, F.W. Billmeyer, Jr.: Small and moderate colour differences: II The Morley data. J. Soc. Dyers Colour. **91**, 229–242 (1975)

7.29 K. McLaren: The Adams-Nickerson color-difference formula. J. Soc. Dyers Colour. **86**, 354–366, 368 (1970)

7.30 K. McLaren, D.A. Plant: "ANLAB – A Uniform Colour Space for Pigment Evaluation", in *Eleventh Congress, FATIPEC*, Florence, 1972 ed. by D. Pagani et al. (Edizioni Ariminum, Milan 1972) pp. 61–66

7.31 K. McLaren: The development of the CIE 1976 ($L^*a^*b^*$) uniform colour space and colour-difference formula. J. Soc. Dyers Colour. **92**, 338–341 (1976)

7.32 K. McLaren, B. Rigg: The SDC recommended colour-difference formula: Change to CIELAB. J. Soc. Dyers Colour. **92**, 337–338 (1976)

7.33 A.R. Robertson: The CIE 1976 color-difference formulae. Color Res. Appl. **2** (1), 7–11 (1977)

7.34 Recommendations on uniform color spaces, color difference equations, and psychometric color terms. Color Res. Appl. **4** (1), 44 (1979)

7.35 D.L. Post: CIELUV/CIELAB and self-luminous displays: Another perspective. Color Res. Appl. **9** (4), 244–245 (1984)

7.36 *Colorimetry*. Supplement No. 2 to CIE Publication No. 15 (CIE, Paris, 1978) [7.42]

7.37 K. McLaren: CIELAB hue-angle anomalies at low tristimulus ratios. Color Res. Appl. **5** (3), 139–143 (1980)

7.38 K.L. Kelly: Twenty-two colors of maximum contrast. Color Eng. **3** (6), 26–27 (1976)

7.39 R.C. Carter, E.C. Carter: High-contrast sets of colors. Appl. Opt. **21**, 2936–2939 (1982)

7.40 A.B.J. Rodrigues, R. Besnoy: What is metamerism? Color Res. Appl. **5** (4), 220–221 (1980)

7.41 *Special Metamerism Index: Change in Illuminant*. Supplement No. 1 to CIE Publication No. 15 (CIE, Paris, 1972) [7.42]

7.42 *Colorimetry*. CIE Publication No. 15 (E-1.3.1) (CIE, Paris 1971)

7.43 J.T. Luke: Induced color demonstrations utilizing colored shadows. Color Res. Appl. **1** (4), 202–205 (1976)

7.44 P.J. Bouma: *Physical Aspects of Colour; An Introduction to the Scientific Study of Colour Stimuli and Colour Sensations*, 2nd ed., ed. by W. de Groot, A.A. Kruithof, J.L. Ouweltjes (Macmillan, London 1971)

7.45 D. Nickerson: Light sources and color rendering. J. Opt. Soc. Am. **50**, 57 (1960)

7.46 D. Nickerson: Terminology of color rendering. J. Opt. Soc. Am. **55**, 213–214 (1965)

7.47 *Method of Measuring and Specifying Colour Rendering Properties of Light Sources*, 2nd ed., CIE Publication No. 13.2 (TC-3.2) (CIE, Paris 1974)

7.48 L. Mori, T. Fuchida: Subjective evaluation of uniform color spaces used for color-rendering specification. Color Res. Appl. **7** (4), 285–293 (1982)

Chapter 8

8.1 H. Pauli: Proposed extension to the CIE recommendation on uniform color spaces, color difference equations, and metric color terms. J. Opt. Soc. Am. **66** (8), 866–867 (1976)

8.2 M.R. Pointer: A comparison of the CIE 1976 colour spaces. Color Res. Appl. **6** (2), 108–117 (1981)

8.3 W. Roddewig: Estimation of chromaticness of constant-luminance central field colors on a D_{65} surround. Color Res. Appl. **9** (1), 49–56 (1984)

8.4 *Chroma Cosmos 5000: Explanatory Note*. (Japan Color Research Institute, Tokyo 1978)

8.5 F. Birren: *Chroma Cosmos 5000*. Color Res. Appl. **4** (3), 171–172 (1979)
8.6 K.L. Kelly: *Chroma Cosmos 5000*. Color Res. Appl. **6** (1), 59–60 (1981)
8.7 *Color Chart System for Design, Chromaton 707* (Japan Color Research Institute, Tokyo 1982)
8.8 F. Birren: *Chromaton 707*. Color Res. Appl. **8** (4), 262 (1983)
8.9 D. Nickerson: Interrelation of color specifications. Pap. Trade J. (TAPPI Section) **125**, 153, 219 (1947)
8.10 W.C. Granville, E. Jacobson: Colorimetric specification of the *Color Harmony Manual* from spectrophotometric measurements. J. Opt. Soc. Am. **34** (7), 382–395 (1944)
8.11 M.E. Bond, D. Nickerson: Color-order systems, Munsell and Ostwald, J. Opt. Soc. Am. **32** (12), 709–719 (1942)
8.12 C.E. Foss, D. Nickerson, W.C. Granville: An analysis of the Ostwald color system. J. Opt. Soc. Am. **34** (7), 361–381 (1944)
8.13 M. Richter: The Official German Standard Color Chart, translated by D.B. Judd, G. Wyszecki, J. Opt. Soc. Am. **45** (3), 223–226 (1955)
8.14 D. Nickerson: Horticultural Colour Chart names with Munsell key. J. Opt. Soc. Am. **47**, 619 (1957)
8.15 The ICI colour atlas. Paint Manuf. **40** (3), 29–30 (1970)
8.16 K. McLaren: Colour specification by visual means. J. Oil Colour Chem. Assoc. **45**, 879–886 (1971)
8.17 G.W. Haupt, J.C. Schleter, K.L. Eckerle: *The Ideal Lovibond Color System for CIE Standard Illuminants A and C Shown in Three Colorimetric Systems*, NBS Note 716 (U.S. Government Printing Office, Washington, D.C. 1972)
8.18 A. Kornerup: The colour system in the *Methuen Handbook of Colour*. J. Oil Colour Chem. Assoc. **47** (12), 955–970 (1964)
8.19 Standard method for specifying color by the Munsell system (ASTM Designation D 1535–62). Off. Dig., Fed. Soc. Coatings Technol. **36** (471), 373–408 (1964)
8.19a *Standard Method of Specifying Color by the Munsell System*. ASTM Designation D 1535–80. (American Society for Testing and Materials, Philadelphia 1980)
8.20 S.M. Newhall, D. Nickerson, D.B. Judd: Final report of the OSA Subcommittee on the spacing of the Munsell colors. J. Opt. Soc. Am. **33** (7), 385–418 (1943)
8.21 D. Nickerson: Spacing of the Munsell colors. Illum.Eng. (N.Y.) **40** (6), 373–386 (1945)
8.22 A. Hård: "Philosophy of the Hering-Johansson Natural Colour System", in *Proceedings of the International Colour Meeting*, Lucerne 1965, Vol. 1 (Musterschmidt, Göttingen 1966) pp. 357–365; Die Farbe **15**, 296 (1966)
8.23 A. Hård: "A New Colour Atlas Based on the Natural Colour System by Hering-Johansson", in *Proceedings of the International Colour Meeting*, Lucerne 1965, Vol. 1 (Musterschmidt, Göttingen 1966) pp. 367–375; Die Farbe **15**, 287 (1966)
8.24 A. Hård: Quality Attributes of Color Perception. Paper presented at *Colour 69*, First Congress of the International Colour Association, Stockholm (1969)
8.25 A. Hård: "The Natural Colour System and Its Universal Application in the Study of Environmental Design", in *Colour for Architecture*, ed. by T. Porter, B. Mikellides (Studio Vista, London 1975) pp. 108–119
8.26 D.B. Judd, D. Nickerson: Relations between Munsell and Swedish Natural Colour System scales. J. Opt. Soc. Am. **65** (1), 85–90 (1975)
8.27 A. Hård, L. Sivik: NCS – Natural Colour System: A Swedish standard for color notation. Color Res. Appl. **6** (3), 129–138 (1981)
8.28 A survey of American color specifications – 1955. Off. Dig., Fed. Soc. Coatings Technol. **28** (381), 902–921 (1956)
8.29 D.L. MacAdam: Colorimetric data for samples of OSA uniform color scales. J. Opt. Soc. Am. **68**, 121–130 (1978); Addenda, 55 pages of tables including tabulations of CIE 1931 (x, y, Y) CIE ILL C for all color samples, *American Institute of Physics (AIP) Document* No. PAPS JOSA-68-121-55 (1978); 570 pages of spectrophotometric and colorimetric data (6 microfiches), *Document* No. PAPS JOSA-69-206-564 (1978)

8.30 D. Nickerson: Munsell renotations for samples of the OSA Uniform Color Scales J. Opt. Soc. Am. **68**, 1343–1347 (1978)

8.31 D. Nickerson: History of the OSA Committee on Uniform Scales. Opt. News **3** (1), 8–17 (1977)

8.32 D. Nickerson: Optical Society of America (OSA) Uniform Color Scale samples. Leonardo **12** (3), 206–212 (1979)

8.33 D. Nickerson: OSA Uniform Color Scale samples, a unique set. Color. Res. Appl. **6** (1), 7–33 (1981

8.34 D. Nickerson: Gleichabständige OSA-Farbreihen, ein einzigartiges Farbmustersortiment (translated by K. Richter). Farbe + Design (12), 16–24 (1979)

8.35 W.E.K. Middleton: The Plochere color system: A descriptive analysis. Can. J. Res. Sect. F **27**, 1 (1949)

8.36 Textile Colors from Munsell. Color Res. Appl. **9** (2), 117–118 (1984)

8.37 G. Reimann, D.B. Judd, J.H. Keegan: Spectrophotometric and colorimetric determination of the TCCA Standard Color Cards. J.Opt. Soc. Am. **36**, 128–159 (1946)

8.38 F. Birren: Color identification and nomenclature: A history. Color Res. Appl. **4** (1), 14–18 (1979)

8.39 *Faber Birren Collection of Books on Color, Art and Architecture Library, Yale University.* A Bibliography. (Yale University Library, New Haven, Conn. 1982)

8.40 R.L.Herbert: A color bibliography: The Faber Birren Collection on color. Yale Univ. Library Gazette **49** (1), 2–49 (1974)

8.41 R.L. Herbert: A color bibliography, II: Additions to the Faber Birren Collection on Color. Yale Univ. Library Gazette **52**, 127–165 (1978)

8.42 The Colour Reference Library – Royal College of Art. Leonardo **17** (1), 50 (1984)

8.43 D. Nickerson, S.M. Newhall: A psychological color solid. J. Opt. Soc. Am. **33** (7), 419–423 (1943)

8.44 *Munsell-CIE (1931) Diagrams.* Set of 14 digarams. (Munsell Color, Baltimore)

8.45 *Munsell Limit Color Cascade.* (Munsell Color, Baltimore 1974)

8.46 W.C. Granville, C.E. Foss, I.H. Godlove: *Color Harmony Manual:* Colorimetric analysis of the third edition. J. Opt. Soc. Am. **40**, 265A (1950)

8.47 W. Schultze: *Farbenlehre und Farbenmessung. Eine kurze Einführung.* 2nd ed. (Springer, Berlin, Heidelberg 1966) (3rd ed. 1975)

8.48 A. Hård: The Swedish Colour Centre Foundation. Color Res. Appl. **11** (1), 8–10 (1986)

8.49 G. Tonnquist: Philosophy of perceived color order systems. Color Res. Appl. **11** (1), 51–55 (1986)

8.50 *NCS, Natural Colour System* (brochure) (Scandinavian Colour Institute, Stockholm 1980)

Chapter 9

9.1 D.L. MacAdam: "System of OSA Committee on Uniform Color Scales", in *AIC Color 77*, ed. by F.W. Billmeyer, Jr., G. Wyszecki. Third Congress of the International Colour Association, Troy, N.Y. (Adam Hilger, Bristol 1978) pp. 399–400

9.2 D.L. MacAdam: Judd's contributions to color metrics and evaluation of color differences. Color Res. Appl. **4** (4), 177–193 (1979)

9.3 D.L. MacAdam: Uniform color scales. J. Opt. Soc. Am. **64**, 1691–1702 (1974)

9.4 G. Wyszecki: A regular rhombohedral lattice sampling of Munsell renotation space. J. Opt. Soc. Am. **44** (9), 725–734 (1954)

9.5 F.W. Billmeyer, Jr.: On the geometry of the OSA Uniform Color Scales Committee space. Color Res. Appl. **6** (1), 34–37 (1981); ibid. **6** (2), 120 (1981)

9.6 C.E. Foss: Space lattice used to sample the color space of the Committee of Uniform Color Scales of the Optical Society of America. J. Opt. Soc. Am. **68**, 1616–1619 (1978)

9.7 *Uniform Color Scales* (brochure accompanying the album of OSA-UCS color samples [1.35]) (Optical Society of America, Washington, D.C. 1977)

Chapter 10

10.1 *ISCC-NBS Centroid Color Charts*. Standard Sample No. 2016. (Office of Standard Reference Materials, National Bureau of Standards, Washington, D.C. 1965)

Chapter 11

11.1 R.W.G. Hunt: A model of colour vision for predicting colour appearance. Color Res. Appl. **7** (2, Pt 1), 95–112 (1982)

11.2 L.M. Hurvich, D. Jameson:Some quantitative aspects of an opponent-colors theory. IV. A psychological color specification system. J. Opt. Soc. Am. **46**, 416–421 (1956)

11.3 D. Jameson, L.M. Hurvich: Theory of brightness and color contrast in human vision. Vision Res. **4**, 135–154 (1964)

11.4 F.W. Campbell: "Adaptation", in *Encyclopaedic Dictionary of Physics*, Vol. I, ed. by J. Thewlis (Pergamon, Oxford 1961)

11.5 D. Jameson, L.M.Hurvich: Some quantitative aspects of an opponent-colors theory. III. Changes in brightness, saturation, and hue with chromatic adaptation. J. Opt. Soc. Am. **46**, 405–415 (1956)

11.6 A.R. Robertson: Critical review of definitions of metamerism. Color Res. Appl. **8** (3), 189–191 (1983)

11.7 H. Helson: "Adaptation Level Theory", in *Psychology: A Study of a Science*, Vol. I, ed. by S. Koch (McGraw-Hill, New York 1959)

11.8 D.B. Judd: Appraisal of Land's work on the two-primary color projections. J. Opt. Soc. Am. **50**, 254–268 (1960)

11.9 M.H. Wilson, R.W. Brocklebank: Colour and perception: The work of Edwin Land in the light of current concepts. Contemp. Phys. **3**, 91–111 (1962)

11.10 Y. Nayatani, K. Takahama, H. Sobagaki, J. Hirono: On exponents of a nonlinear model of chromatic adaptation. Color Res. Appl. **7** (1), 34–45 (1982)

11.11 C.J. Bartleson: Memory colors of familiar objects. J. Opt. Soc. Am. **50** (1), 73–77 (1960)

11.12 G.S. Brindley: Afterimages. Sci. Am. **209** (10), 84–91 (1963)

11.13 S. Anstis, B. Rogers, J. Henry: Interactions between simultaneous contrast and colored afterimages. Vision Res. **18**, 899–911 (1978)

11.14 J.P.C. Southall (ed.): *Helmholtz's Treatise on Physiological Optics*, Vol. II, translated from 3rd German edition (Dover, New York 1962)

11.15 I.T. Pitt, L.M. Winter: Effect of surround on perceived saturation. J. Opt. Soc. Am. **64**, 1328–1331 (1974)

11.16 D. Jameson, L.M. Hurvich: Opponent chromatic induction: Experimental evaluation and theoretical account. J. Opt. Soc. Am. **51** (4), 46–53 (1961)

11.17 E.H. Land: Color vision and the natural image. Proc. Nat. Acad. Sci. USA **45**, 115–129, 636–644 (1959)

11.18 E.H. Land: Experiments in color vision. Sci. Am. **200** (5), 84–99 (1959)

11.19 C.S. McCamy: A demonstration of color perception with abridged color-projection systems. Photogr. Sci. Eng. **4**, 155–159 (1960)

11.20 A. Karp: Colour-image synthesis with two unorthodox primaries. Nature (London) **184**, 710–712 (1959)

11.21 R. Belsey: Color perception and the Land two-color projections. J. Opt. Soc. Am. **54** (4), 529–531 (1964)

11.22 A. Fiorentini: "Mach Band Phenomena", in *Handbook of Sensory Physiology*, Vol VII/4, *Visual Psychophysics*, ed. by D. Jameson, L.M. Hurvich (Springer, Berlin, Heidelberg 1972) pp. 188–201

11.23 D. Jameson: "Opponent Colors Theory in the Light of Physiological Findings", in *Central and Peripheral Mechanisms of Color Vision*. Proceedings of symposium, Stockholm June 14–15, 1984 (Macmillan, London 1985)

Chapter 12

12.1   Manufacturers Council on Color and Appearance (MCCA): Guide to educational opportunities, publications, and associations dealing with color and appearance technology. Color Res. Appl. **5** (4), 222–227 (1980)

12.2   Munsell Color Science Laboratory funded. Color Res. Appl. **8** (3), 196–197 (1983)

12.3   K.L.Kelly, D.B.Judd: "The ISCC-NBS Method of Designating Colors and a Dictionary of Color Names", in [7.2]

12.4   C.J. Bartleson: Changes in color appearance with variations in chromatic adaptation. Color Res. Appl. **4** (3), 119–138 (1979)

12.5   J.M. Eichengreen: *Time Dependent Chromatic Adaptation*, Ph.D. dissertation (University of Pennsylvania, Philadelphia 1971)

Name and Subject Index

Tristimulus values (CIE) *see* CIE, tristimulus values: X, Y, Z; X_{10}, Y_{10}, Z_{10}

Tungsten-filament lamps *see* Incandescent lamps, light from

Turner, J.M.W. 1

Twilight vision *see* Mesopic vision

Two-color projections 205, 206

Ultraviolet radiation *see also* Black light 17, 18, 23, 37, 38, 39, 40, 239

Uniform color arrays, OSA *see* OSA, uniform color arrays

Uniform color scales, OSA *see* OSA, uniform color scales

Unique hues *see* Hues, unitary

Unitary hues (Unique hues) *see* Hues, unitary

Unrelated colors *see* Isolated colors

U.S. National Bureau of Standards *see* National Bureau of Standards (NBS) (USA)

Value 16

Value, Munsell *see* Munsell, Value; Munsell, units

Varnished paintings 34

Vasarely, V. 208

Vibration 94

Villalobos Color Atlas 113

Visible spectrum 17, 18, 19, 20

Vision, color (photopic vision) 5, 8, 27, 40, 41, 43, 188, 190, 191, 195

Visual angle *see* Angle, of vision

Visual mixture *see* Color mixture, by averaging

Wave frequency 18

Wavelength 18, 19, 37
  composition 10, 24, 35
  intervals (bands) 20
  measure 18, 59, 60, 63, 66

Wehlte, K. 28

White 13, 29, 78, 241, 242

White, equal-energy *see* Equal-energy white

White light 41, 42, 72, 97, 99

Whiteness, NCS *see* NCS, whiteness

Wilson, M.H. 197, 198, 199, 248

Wratten light filters 98, 113

Wright, W.D. 16, 53, 65, 193, 244

Wyszecki, G. 2

Springer Series in Optical Sciences

Editorial Board: D.L. MacAdam A.L. Schawlow K. Shimoda A.E. Siegman T. Tamir